Lexicon of Biological Sciences
Vol. : 1 Forestry

Lexicon of Biological Sciences

Vol. : 1 Forestry

K. VANANGAMUDI
Dean (Agriculture)
Tamil Nadu Agricultural University
Combatore, Tamil Nadu

K. NATARAJAN
Asstt. Professor of Seed Science & Technology

K.R.V. SATHYA SHEELA
Asstt. Professor of Plant Breeding & Genetics

Adhiparasakti Agricultural College, Kalavai
Vellore District - Tamil Nadu

2008

SSPH

SATISH SERIAL PUBLISHING HOUSE
403, Express Tower, Commercial Complex, Azadpur, Delhi-110033 (India)
Phone : 011-27672469, 27672852, Fax : 91-11-27672046
E-mail :info@satishserial.com, hkjain1975@yahoo.com
Website : www.satishserial.com

Published by :

SATISH SERIAL PUBLISHING HOUSE
403, Express Tower, Commercial Complex, Azadpur, Delhi - 110033 (INDIA)
Phone : 011-27672852 Fax : 91-11-27672046
E-mail : info@satishserial.com, hkjain1975@yahoo.com

ISBN : 81-89304-38-0

Printed at
SALASAR IMAGING SYSTEMS, DELHI-35

PREFACE

Forests have been distributed all over India since time immemorial. Forestry as a formal sector came into existence only during colonial period. India is one of the first countries to have begun the scientific management of forests earlier than any other countries. Forest industry as defined includes all establishments that manage timber products harvest and transport timber logs and process logs into products. Forests contribute significantly to the economic basis of local communities. Forest are maintained and managed to meet the needs of defence, communications and industry and other general purposes of public like fuel wood and timber. Moreover forests must be preserved or created for physical and climatic consideration.

Forestry is a broad subject which deals on multi-disciplinary aspects like silviculture, forest soils, properties of soils, forests policies, forest law, botany of tree species, surveying protection, logging which includes felling and conversion, transport, storage, mensuration, grading timber *etc*. There is no horizon end for the subject forestry, in which there is tremendous progress in recent years. The role of forestry in not just limited to economic rewards but enhances the quality of life making the place more attractive to live and work.

The coverage of terms is very broad and it is very hard to think more cherished source of information in forestry at a glance. Hence an attempt has been made to bring out a comprehensive glossary of technical terms used in the theory and practice of forestry. Each definition is considered a short lesson to be understandable as an independent entry.

Moreover diagrammatic representations are also included in each subject for easy understanding of the terms. The aim of the glossary is to provide a concise and enough information and examples to give the reader a sound knowledge of forestry.

The glossary is made in such a way that it caters to all kinds of audiences, be it an amateur in the field or a veteran in the field and helps them in teaching and learning of forestry. Sincere efforts have been taken to make the readers immerse in the wonderful discipline of forestry.

Authors

CONTENT

A

Abiotic

A phenomenon that refers to the absence of living organisms.

Abney Level

An manual surveying instrument designed for the measurement of angles of elevation or depression. It is expressed in degrees or percent.

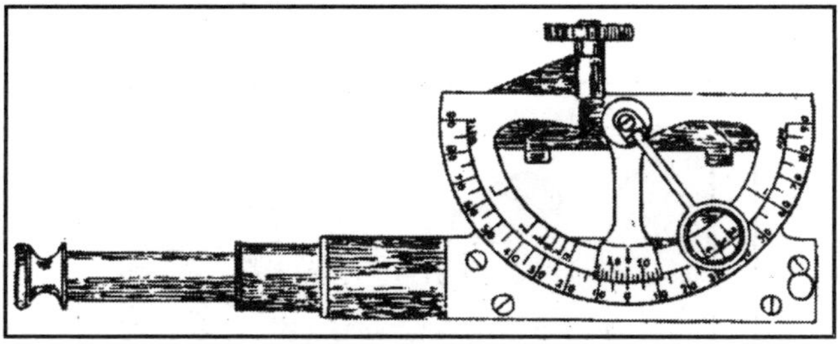

Abneys level

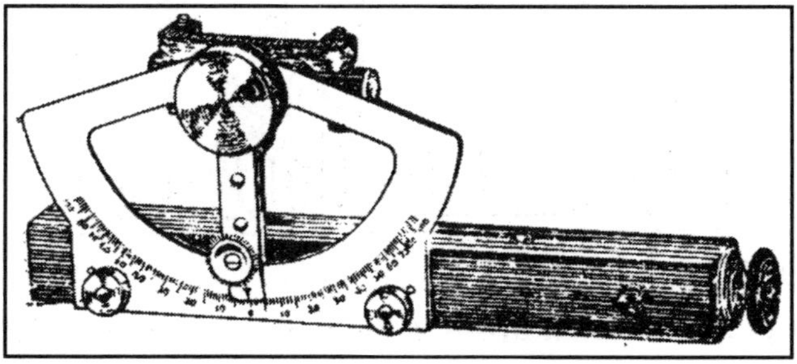

Topographical abneys level

Acre

An unit of land which measures 4000 sq m or 43,560 sq ft or 10 square chains, or a square 208.7 ft on each side.

Adaptation

The adjustment of an organism to new or altered environmental conditions by the occurrence of genetic changes in a population or species as the result of natural selection.

Advance Growth

The young aged trees that have become established naturally in a forest before beginning of cutting or regeneration.

Advance Regeneration

It refers to the young trees under existing stands i.e., regeneration established before logging that has survived the logging operation.

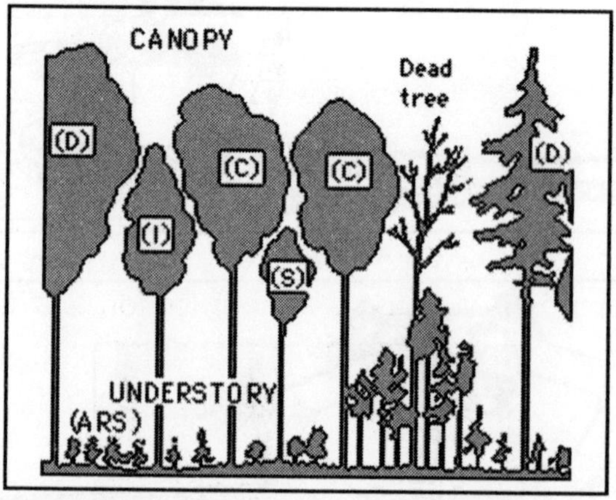

Adventitious

A plant part that arises outside the usual order of time and/or position, e.g., an adventitious bud develops from any part of a stem, leaf, or root but lacks vascular connection with the pith; an adventitious root develops from parts of the plant other than a pre-existing root, e.g., from a stem or leaf.

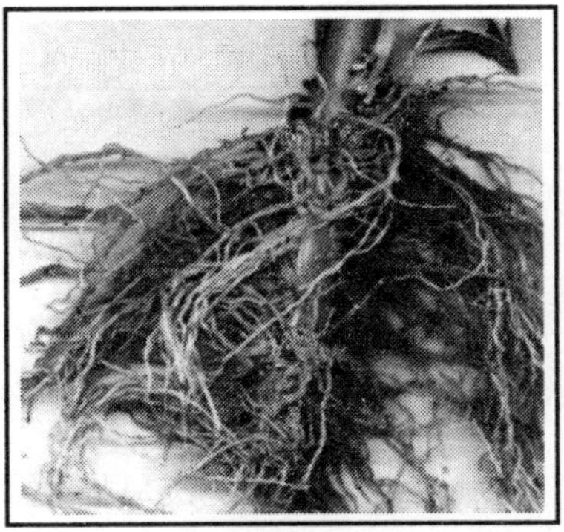

Afforestation

Afforestation refers to the establishment of a forest on an area that has not been previously forested.

Age Class

Age class is one of the intervals into which the age range of forest stands is divided for classification and use.

Agro-Forestry

Agro-forestry defined as the practice of raising trees, forage, and livestock on the same ground, at the same time which includes common associations like are cattle and trees or sheep.

All-aged

All-aged is a stand that contains trees of all ages.

All-aged Management

It is a system of growing forest trees in groups where the individual trees are not the same age. An all-aged forest is a type of forest that has trees scattered throughout a range of age from I year to the oldest tree, whatever its age may give a space be.

All-Aged Structure

The term applies to a stand that comprises trees of most or all age classes, from seedlings to mature trees.

Allegany Hardwood Forest Type

It is a portion of Maryland's northern hardwood forest in which black cherry, white ash, and red oak are dominant species.

Allelopathy

The phenomenon of the influence of plants upon each other, through the products of their metabolic reactions.

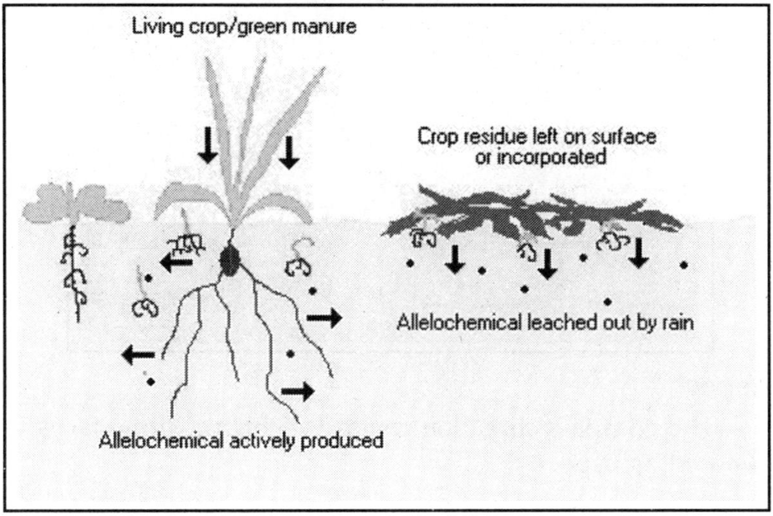

Allowable Cut

It is the volume of wood that may be harvested, under management, for a given period.

Annual Ring

It refers to the growth layer of I year, as viewed on the cross section of a stem, branch, or root. One year's growth consists of a layer of lighter-colored wood (springwood) and a layer of darker-colored wood (summerwood).

Arch

An inverted V or U shaped trailer or structure, which is used in logging and is towed behind (or attached to the back of) the skidding machine, lifting one end of the logs off of the ground during the yarding operation.

Aspect

The compass direction towards which a slope faces is denoted as aspect.

Available Water

Available water is defined as the amount of moisture in the soil that can be used by the plants, usually which does not includes water that drains readily and water beyond the "wilting point".

Area Ignition

The term refers to the practice of setting a number of individual fires throughout an area, either simultaneously or in quick succession, and so spaced that they soon coalesce, influence, and support each other to produce a hot, fast spreading fire throughout the area.

Artificial Regeneration

A practice of direct seeding or planting seedlings or cuttings for the renewal of a tree crop.

Association

Association is a collection of plants with ecologically similar requirements, including one or more dominant species from which the group derives a definite character.

—xxx—

B

Bacillus thuringiensis

A bacterial microorganism that is used for producing, biological insecticide that kills spruce budworm, gypsy moths, forest tent caterpillars, and other insects.

Back Cut

The practice of the final cut, made on the opposite side of the tree from the face cut (or undercut) in the process of felling a tree.

Backfire

The controlled fire set ahead of a forest fire for creating a firebreak by reducing fuel in the path of the main fire.

Backlog

Backlog is an administrative term used to classify inadequately stocked forest land that has been denuded (cut over, burned, etc.).

Basal Area

Basal area of a tree refers to the cross-sectional area (in m^2) of a trunk measured at breast height of a tree ie., at 1.4 m above the ground. eg. the basal area of a tree 40 cm dbh is approximately 0.5 m^2. Basal area of a hectare of forest refers to the sum of basal areas of individual trees on the hectare. Eg. a well stocked hardwood stand might contain 18 to 23 m^2 of basal area.

Basal Bark Treatment

It is a treatment for killing trees and brush in which a herbicide is applied, by sprayer or brush, to a band of bark encircling the basal portion of the stem.

Basic Forest

Basic forest is the extensive forest management plus artificial regeneration where necessary. cf. extensive forest management

Basic Silviculture

It is the practice of silviculture required to achieve a free-growing (or established) regeneration of desired species at specified densities and stocking.

Bedding

It is a procedure associated with site preparation in which the soil is mounded mechanically to provide a well-drained ridge on which seedlings are planted.

Bench Mark

Benchmark is a point of known elevation usually referenced to sea level.

Biltmore Stick

Biltmore sticks are tools calibrated with different scales depending on the users' arm length to measure the diameter of a tree at breast height

Bioassay

A test of a particular unknown variable by using the performance of a living organism as an indicator.

Biological Control

A method of control of plants, diseases, and animal pests by the use of natural enemies.

Biological Diversity or Biodiversity

Biodiversity refers to diversity of genetics, species, ecosystems, and landscapes which includes variety of life in all its forms and all its levels of organization.

Biomass

Refers to the sum total of biological material that exists on a given land area. For trees and plants biomass comprises leaves, branches, stems, and roots.

Biome

It refers to an ecological community of plants and animals extended over a large natural area.

Biosphere

Biosphere is the life zone of the earth that includes the lower part of the atmosphere, the hydrosphere, soil, and the lithosphere to a depth of about 2 km.

Biotic

The term pertains to life and living organisms.

Blowdown

The term refers to the trees that are knocked down by wind.

Blue stain

The fungus discoloration common in pines, predominantly bluish, but sometimes greyish, blackish, or brownish in appearence and confined almost exclusively to sapwood.

Blazing

The process of marking a tree, usually by painting or cutting the bark. Forest properties are delineated by blazing trees along the boundary lines.

Block Cutting

It is the practice of removal of the crop in blocks in one or more operations, generally for wildlife management purposes, encouraging regeneration, or protecting fragile sites.

Board Foot

A unit for measurement of wood volume in a tree, log, or board which is commonly 1 foot by 1 foot by 1 inch. But any shape containing 144 cubic inches of wood equals one board foot.

Bole

Refers to the main trunk of a tree.

Bolt

A short log or a squared timber cut from a log.

Boomer

Boomer is a slang name for mountain beaver.

Boots

Boots are used by firefighters to protect their feet on the fireline and rugged terrain.

Boreal Forest Region

It is the largest forested area in Saskatchewan, dominated by conifer species like spruce, fir, pine and tamarack. A mixture of deciduous trees such as birch and poplar is found in the southern edge of this region. This region forms a band extending diagonally across the northern half of the province.

Borrow Pit

An area in roadbuilding where fill material is "borrowed," and used in road sections where the normal excavation process does not generate adequate fill material.

Breast Height Age

The term refers to the number of annual growth rings between the bark and the pith counted at breast height.

Breast Height

Breast height is the standard height above ground level at which the diameter of a standing tree is measured

(1.3 m). The breast height is usually measured on the uphill side of the tree for a sloping ground.

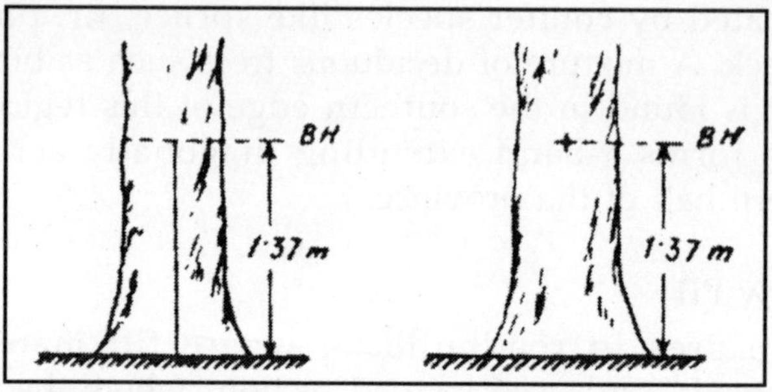

Breast height should be marked by means of a measuring stick on standing trees

Breeding Orchard

A planting of selected trees, usually clonally propagated, designed to ease breeding work. It is also called breeding seed orchard and breeding arboretum.

Breeding Population

A group of individuals selected from a wild population for use in a breeding program selected phenotypically for desirable traits. There are usually several to many, more-or-less separate breeding populations, each designed to provide progeny suited to a particular geographic region in species with large range.

Breeding Value

The genetic value of an individual determined by the mean value of its progeny on the basis of individual traits or a selection index.

Breeding Zone

It is an area within which a single population of improved trees can be planted without fear of mis-adaptation.

Broadcast Burn

Broadcast burn is a controlled burn in which the fire is allowed to proceed over an entire area. Sometimes called a slash burn.

Brush

Commonly refers to undesirable shrubs and other low lying vegetation.

Budburst

It denotes the time in the spring when flower or leaf buds begin their annual growth in woody plants.

Broadleaf

A tree with flat leaves and flowers that produce fruit when developed and fertilized. The group comprises all deciduous trees and a very few ever greens, such as holly.

Browse

Browse denotes the parts of woody plants like twigs, shoots, and leaves which are eaten by forest animals. It also includes small bushes, sprouts, herbaceous plants and small trees.

Brush

It applies to the shrubs and stands of short, scrubby tree species that do not reach merchantable size.

Brushing

Brushing is a practice of the removal of undesirable herbaceous and woody vegetation by mechanical means.

Bucking

The process of cutting trees into shorter lengths, such as logs or cordwood.

Bud Pruning

Removal of lateral buds from a stem to prevent them from developing into branches.

Bud cap

Bud cap is a piece of paper or other suitable material attached to a young seedling, covering the terminal bud for preventing animal browsing.

Buffer

Buffer is a zone or strip of land that shields one area from another commonly used along streams or as visual barriers.

Bug Kill

Tree or timber stands killed by insects.

Bunk

It refers to support on a railroad car or logging truck on which the logs rest. ie., a bed in a logging camp.

Burl

Burl is a abnormal growth on a tree stem, with wood tissue growing in an irregular pattern. Usually circular in shape, these growths are widely sought for their interesting grain pattern.

Burn and Controlled

Any type of burning that is started intentionally to accomplish a particular purpose, and having some surveillance or control over the process.

Burn and Prescribed

The application of fire to land under conditions of weather, soil moisture, and time of day, that will accomplish specific silvicultural, wildlife, grazing, or fire hazard reduction purposes.

Butt Log

Butt log is a log cut from the bole immediately above the stump.

Butt

The base of a tree or log.

—xxx—

Burl

Burl is a abnormal growth on a tree stem, with wood tissue growing in an irregular pattern. Usually circular in shape, these growths are widely sought for their interesting grain pattern.

Burn and Controlled

Any type of burning that is started intentionally to accomplish a particular purpose, and having some surveillance or control over the process.

Burn and Prescribed

The application of fire to land under conditions of weather, soil moisture and time of day that will accomplish specific silvicultural, wildlife, grazing, or fire hazard reduction purposes.

Butt Log

Butt log is a log cut from the bole immediately above the stump.

Butt

The base of a tree or log.

--xxx--

C

Caliper (or calipers)

An instrument used to measure diameter of trees or logs. It consists of two parallel arms at right angles to a graduated rule, with one arm that slides along the rule.

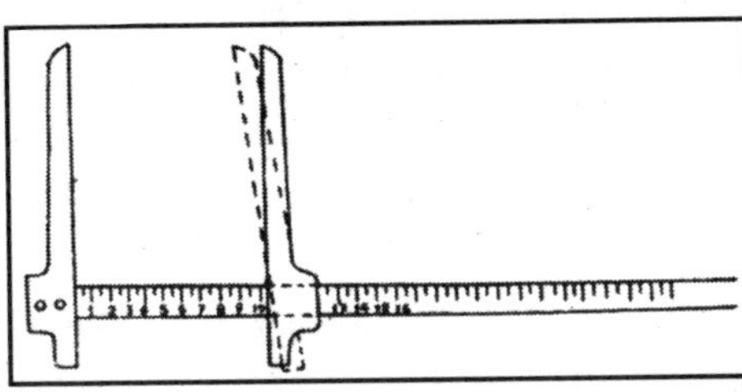

Caliper

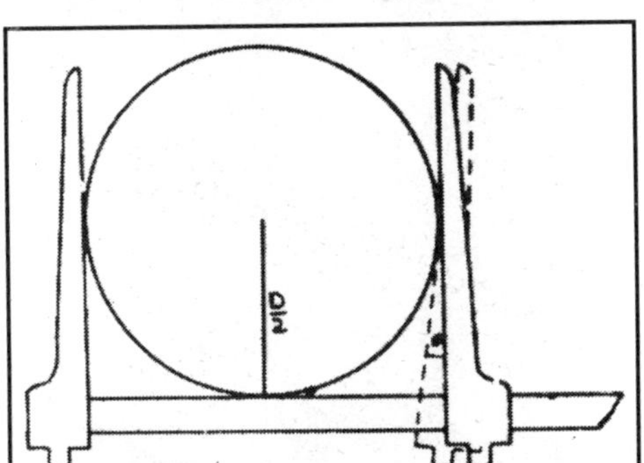

Error in caliper measurement due to wrong orientation of axis

Cambium

Refers to a layer of cells between the woody part of the tree and the bark and the division of these cells results in diameter growth of the tree through formation of wood cells (xylem) and inner bark (phloem).

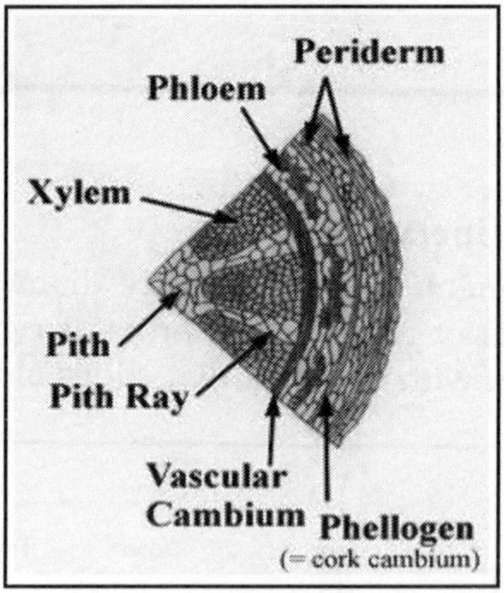

Canopy

A collective term for the layer formed by the crowns of the taller trees in a forest.

Candidate Tree

A tree that has been tentatively selected for inclusion in a breeding program, but has not yet been measured or compared with surrounding trees.

Canopy

Canopy refers to the continuous cover of branches and foliage formed collectively by the crowns of adjacent trees.

Canopy Closure

The reduction of space progressively between crowns as they spread laterally, increasing canopy cover.

Canopy Density

It refers to the amount of foliage cover, combining the extent of canopy closure and crown density.

Capillary Action

Capillary action is the tendency of water to move in fine spaces due to the molecular attraction between the water molecules and the vessel walls.

Carrying Capacity

Carrying capacity is the maximum number of individuals of a wildlife species that an area can support during the most unfavorable time of the year.

Catface

A scar on the surface of a log.

Cellulose

A carbohydrate (CHO) that forms the walls of plant cells and makes up the bulk of the wood in trees.

Character (Trait)

Character means a distinctive but not necessarily invariable feature that is exhibited by all individuals of a group. Eg. color, size, performance. A character of an individual is determined by the individual's genotype and environment and expressed as phenotype.

Check

The stagnation of tree or stand growth is called check.

Chromosome Number

The number of chromosomes present in a typical individual of a particular species. The sex cells of a species have one-half the number of chromosomes found in the vegetative portion and are said to be haploid.

Clay

A soil containing more than 40% clay, less than 45% sand and less than 40% silt consisting of particles smaller than 0.002 mm in diameter.

Clearcut

It refers to either an area of forest land from which all merchantable trees have recently been harvested or the process of the harvesting of all merchantable trees from an area of forest land.

Clearcut

The process of harvesting of all the trees in an area to aid the species whose seedlings require full sunlight to grow well is called clearcut.

Clearcutting Method

A method of harvesting trees for regenerating an even-aged forest stand in which new seedlings become established in fully exposed micro environments after removal of most or all of the existing trees. It is often used in species like jack, pine, where clearcutting is practiced in blocks, strips or patches.

Climax Ecosystem

It is the semi-permanent culminating stage of succession which continues to occupy an area unless there are significant changes in the environment.

Cline

Cline refers to an environmental gradient (temperature, rainfall, soil pH.) and a corresponding phenotypic gradient in a population of plants or animals. Clines are evaluated by provenance tests and they have a genetic basis.

Clinometer

An instrument used to determine the height of a tree.

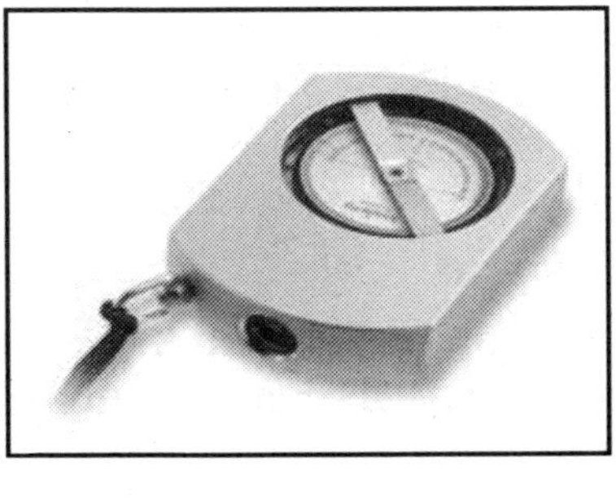

Clonal Test

The field planting of several to many vegetatively propagated plants to estimate the relative performance of different genotypes. They do not necessarily provide information on breeding behavior.

Clone

An asexual propagation of plants by grafting, rooting cuttings, tissue culture, or apomictic seed. Unless there is low level of mutation, all plants from a clone are genetically identical. In tissue culture it means to regenerate a whole plant from a single cell. Whereas in molecular biology it refers to identify and isolate a gene controlling a specific trait from an organism. Generally it refers to a group of plants produced from cuttings, stump or root sprouts, tissue culture, or some other method that produces offspring genetically identical to the original plant. Very often it is used in forestry to establish grafted seed orchards, and it is becoming more common for commercial plantations.

Codominant

Trees with crowns forming the general level of the main canopy in even-aged groups of trees which receive full light from above and comparatively little from the sides.

Combining Ability

It is defined as the ability of an individual to produce progeny with strong expression of a particular trait on crossing.

Commercial Clearcut

Commercial clearcut is a harvest cut that is done to remove all merchantable timber from the area.

Commercial Forestland

Any area capable of producing 20 cubic feet of timber per acre per year that has not been protected from such use by law or statute is refered as commercial forestland.

Commercial Treatments

The term applies to timber stand improvements like thinning which can generate income from the sale of the trees removed.

Community

The collection of living organisms thriving in an organized system through which water, energy, and nutrients cycle is called as community.

Competition Control

A treatment done for reducing the competitive effect of undesirable vegetation threatening the success of regeneration.

Composition

It refers to the proportion of each tree species in a stand expressed as a percentage of either the total number, basal area, or volume of all tree species in the stand.

Cone Rake

Cone rake is a device used for the collection of cones from a standing tree. Cones or cone bearing branches are removed and retrieved by the machine by lowering the machine from the helicopter over the crown of a tree.

Conelet

The immature cone (strobilus) in the conifers is called as conelet.

Conical

Conical refers to the growth of the terminal bud grows much more quickly than the lateral branches to form a cone shape.

Conifer

Any tree that produces seeds in cones is a conifer.

Conservation

Conservation is the wise use of natural renewable resources to ensure their continuing quality and availability to current and future operations.

Consulting Forester

An independent professional who manages forests and markets forest products for private woodland owners. Consulting foresters do not have direct connections with firms that buy wood products, but are retained by woodland owners as their agents.

Control Pollination

Pollination of the female flowers of a tree with pollen from a known source, usually one specific tree. To prevent the female flowers from undesirable pollen, the flowers are covered with cloth or paper bag before they are receptive.

Coppice

It is a natural regeneration that originates from stump sprouts or root suckers.

Coppice Forest

It is a type of forest originating from sprouts or suckers.

Coppice Method

The method of regeneration of a forest stand where the cut trees produce sprouts or suckers.

Coppice-With-Standards Method

It applies to a method of regeneration of a forest stand by coppicing. The selected trees, which may be from seed, are left to grow to larger in size than the coppice beneath them.

Cord

A unit of wood cut for fuel that is equal to a stack 4 x 4 by 8 feet or 128 cubic feet. A cord is the legal measure of fuelwood volume in Maryland.

Cord

Cord is a volume measure of stacked wood. A standard cord is 4 x 4 x 8 ft or 128 cu ft of space. Since roundwood cannot be stacked to give solid volume, actual wood volume varies between 70 and 90 cu ft per cord.

Cordwood

It is a small diameter or low quality wood suitable for firewood, pulp, or chips but not suitable for sawlogs.

Cover Crop

A suitable leguminous herbaceous crop also including crops of cruciferae and gramineae, grown to increase the fertility of soil. The cover crop can be dug or ploughed in while succulent, with or without supplementary fertilisers.

Critical Area

Land in Maryland which is situated within 1,000 feet of the Chesapeake Bay and its tributaries and is subject to forestry and other land use regulations.

Crook

A tree defect characterized by a sharp bend in the main stem.

Crook

The defect in logs and poles or piling, consisting of an abrupt bend or an edgewise warp in a piece of lumber.

Crop Planning

The process of custom designing the density of regeneration and the timing and intensity of stand-tending treatments for achieving site- and species-specific stand management objectives as well as to attain forest-level management objectives.

Crop Tree

The tree that is identified to be grown until the final harvest cut which is usually selected on the basis of its location with respect to other trees based on its quality, rate of growth, species, and straightness.

Cross

Collection of pollen from one tree and pollinating a second tree. Cross also refers to a progeny of a control pollination.

Cross Section

The section of a stem or leaf taken at right angles to its longitudinal axis is called cross section.

Crossing Groups

The group of individuals that comprises a unique set of parents. Eg. A diallel CG has controlled crosses that are made between each pair of parents in the group but crosses with parents outside the group are excluded. A factorial CG includes only a limited number of parents that are used as male testers in controlled crosses with an unlimited number of female parents. An open-pollinated CG has all parents in a breeding population that are included in a progeny test or series of tests. It can be used for reforestation with genetically improved material.

Cross-pollination

Cross pollination refers to the pollination by genetically different plants

Crosstie

A square timber used for supporting railroad rails.

Crotch

It refers to the fork of a tree or branch.

Crown Class

It is the designation of trees in a forest with crowns of similar development occupying similar positions in the crown cover. Differentiation into crown classes applies to even-aged stands and within small even-aged groups in which trees in an uneven-aged stand are often arranged. Dominant, codominant, intermediate, overtopped (suppressed), and wolf trees are the five crown classes which are commonly recognized.

Crown Closure Class

Any interval into which the range of proportions of ground area covered by the vertically projected tree crown areas of a stand is divided for classification and use is called as crown closure class.

Crown Closure

Refers to the time at which the available crown space has become fully occupied.

Crown Cover

Crown cover refers to the canopy of green leaves and branches formed by the crowns of all trees in a forest which is expressed as a percent of total area. crown density. The compactness of the crown cover of the forest depends on the distance apart and the compactness of the individual crowns. It combines the meanings of "crown closure" and "shade density."

Crown Fire

A fire that runs through the tops of living trees, brush, or chaparral.

Crown Cover or Crown Closure

Refers to the percentage of a given area covered by tree crowns.

Crown

Crown refers to the upper part of a tree or other woody plant, carrying the main branch system and foliage.

Crown Density

Crown density applies to the amount, compactness, or depth of foliage of a tree crown.

Crown Ratio or Live-Crown Ratio

The ratio of the leaved portion of a tree's height to its total height is called crown ratio.

Crown Thinning

The removal of trees from the middle and upper crown classes in a stand, to favor the most promising trees of these classes.

Cruise

The term refers to forest survey used to obtain inventory information and develop a management plan.

Cruising

Measuring standing trees to determine the volume of wood on a given tract of land, which can be used for harvesting, purchasing, and general management.

Crummy

Crummy is a crew bus that transports loggers or other woodsworkers to and from the woods.

Crushing

Crushing is the compaction of slash and brush by machinery. The two important features of crushing are slash and provision of microsites.

Cubic Foot

One cu ft of wood in a log usually produces from 3 to 10 board ft of lumber because of the cylindrical log shape and sawing losses.

Cull

A tree or log of merchantable size with poor form, large limbs, rot, or other defects and is rendered unmerchantable.

Cull Tree

A live tree of merchantable size but unmerchantable because of defects or decay.

Cunit

A measurement equal to 100 cu ft of solid wood which is often used for pulpwood measurement.

Cutting Area

It is a portion of woodland on which timber is being cut or will be cut.

Cutting Cycle

The planned interval between major harvesting operations in the same stand. Eg. A 10 year cutting cycle indicates thinnings done once every 10 years.

—xxx—

D

Deciduous

Broad leaved trees that drop all their leaves sometime during the year particularly during autumn.

Debarker

Debarker is a machinery that removes bark from a log.

Deck

Deck refers to stacks of logs.

Decomposer

Decomposer is a heterotrophic organism (including bacteria and fungi) which breaks down the complex compounds of dead protoplasm by absorbing some decomposition products, and releasing substances usable by consumers.

Defect

The term applies to the irregularity or imperfection in a tree, log, piece, product, or lumber which reduces the volume of sound wood or lowers its durability, strength, or utility value.

Defoliation

The process of removal of leaves or needles from plants.

Den Tree

Den tree refers to a tree with cavities suitable for birds or mammals to nest in.

Dendrochronology

It is the study of tree growth rings for establishing dates, environmental conditions, and influences on growth in the past.

Dendrology

Dendrology is defined as the study of the identification and classification of trees.

Diameter

The longest distance at right angles, across any circle or cylinder. The diameter is estimated in standing trees by dividing the circumference by 3.1416.

Diameter Class

It refers to the intervals into which the range of diameters of trees in a forest is divided for purposes of classification and use. Generally this is done in 12 in, even increments (12 in class would contain trees from 11 to 13 in).

Diameter Limit

Diameter limit is the smallest (occasionally the largest), size to which trees or logs are to be measured, cut, or used and the limit usually refers to stump, breast height, or top.

Diameter Breast Height (DBH)

The stem diameter of a tree measured at breast height (1.3 m above ground level) that generally applies to the outside bark dimension.

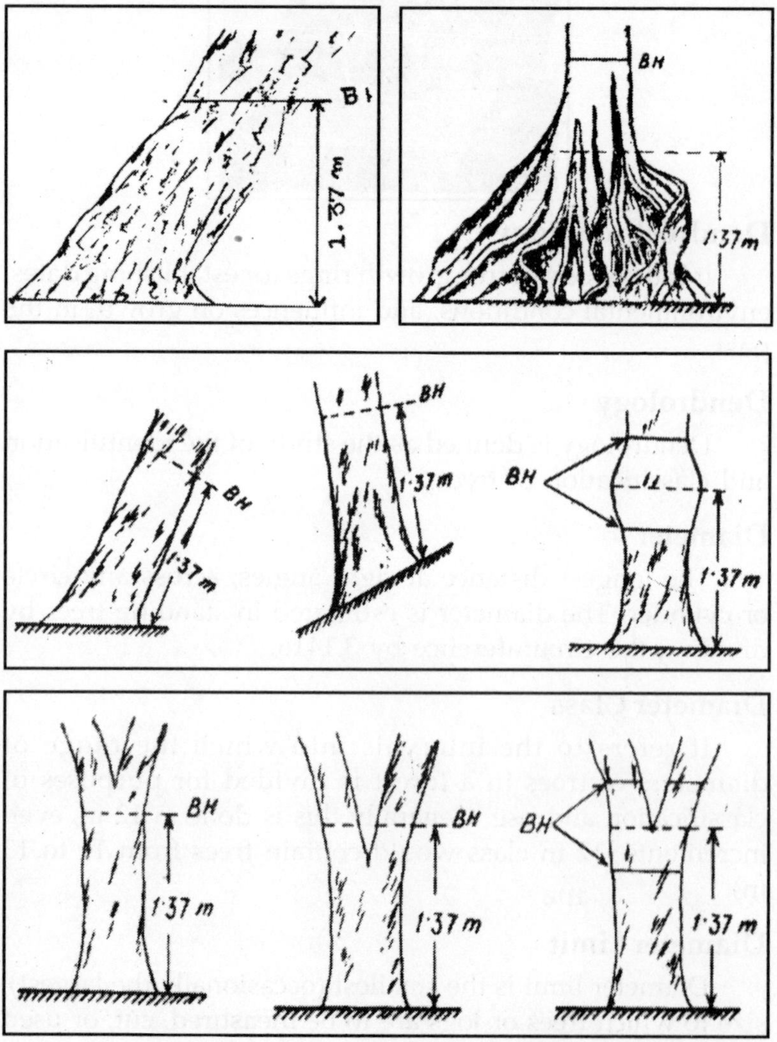

Showing the various measurement of diameter breast height

Diameter Inside Bark (DIB)

The diameter of a tree or log excluding double bark thickness.

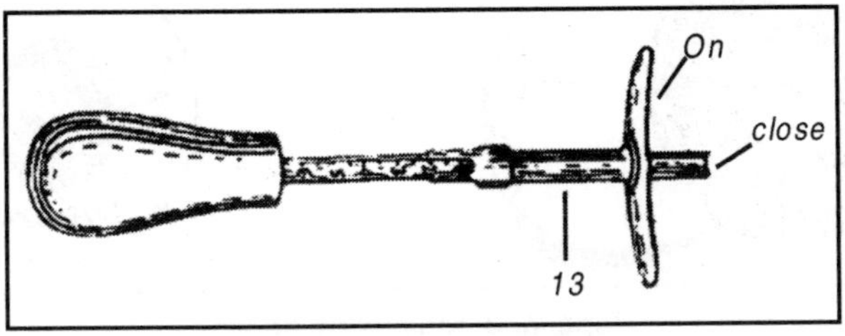

Swedish bark gauge

Diameter-Limit Cutting

The removal of all merchantable trees above a specified minimum diameter, which in mixed stands may vary with species.

Diameter Outside Bark (DOB)

The diameter of a tree or log including bark.

Diameter Over Stubs (DOS)

The outer diameter of the knotty core in pruned trees.

Diameter Stump Height (DSH)

The stem diameter of a tree measured at stump height.

Diameter-Limit Scale

Diameter limit scale is a timber scale in which all trees over a specified dbh may be cut. It often results in high grading.

Diameter Tape

A graduated tape based on the relationship of circumference to diameter, which provides a direct measure of tree diameter when stretched around the outside of the tree.

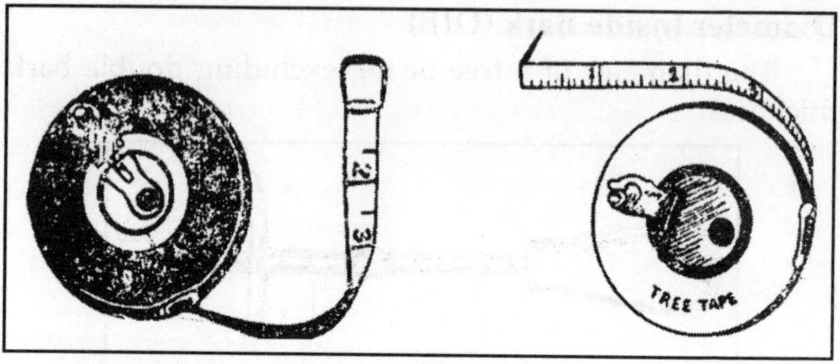

Tape and tape with hook

Dibble

Dibble is a flat or round metal tool used to make a hole for planting containerized seedlings.

Dieback

To die from the top or peripheral parts.

Dimension Lumber

Hardwood dimension lumber is processed to be used whole in the manufacture of furniture or other products. Softwood dimension lumber consists of boards more than 2 inches thick but less than 5 inches thick and is sold as 2 by 4s, 4 by 8s, or 2 by 10s and used in construction.

Direct Seeding

Method of sowing tree seed directly to regenerate a forest.

Disease

The term refers to the harmful deviation from normal functioning of physiological processes, generally pathogenic or environmental in origin.

Dominant Crowns

The uppermost branches of trees that can reach above their neighbours to get more sunlight is called dominant crown.

Dominant Species

The species that is numerous and vigorous in a mixed crop.

Dominant Trees

Refers to trees with crowns extended above the general level of the crown cover and receive full light from above and partly from the side. They are larger than the average trees in the stand, with crowns well developed, possibly somewhat crowded on the sides.

Douglas-fir Region

The region from northern California to southern British Columbia, west of the Cascades crest and the western portions of Oregon and Washington, where Douglas fir grows in uneven aged stands and it is the predominant timber species.

Dry Rot

A type of decay or brown rot caused by specialized fungi which is capable of conducting moisture from an available source and extending their attack to wood previously too dry to decay. There is a misinterpretation that wood will rot when dry, which is not true.

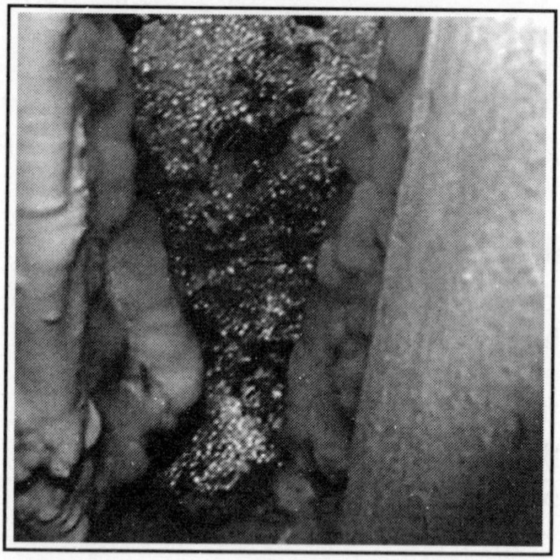

D-plus (D +) Rule

A thumb rule followed for thinning which is obtained by adding a given number to the d. b. h. of the crop tree. Eg. "D + 4" rule would mean that a 16in d.b.h. tree would need 16 + 4 or 20 ft of growing space.

Dormancy

Dormancy is a biological process in which a plant ceases most growth activities due to periods of moisture and/or temperature stress and simply maintains existing tissue.

Drip Torch

It is used to drip a flaming liquid mixture of diesel fuel and gasoline, to ignite fires in burnout operations by the firefighters.

Duff

It is forest litter and other organic debris in various stages of decomposition on top of the mineral soil, where the decomposition rate is slower and litter accumulation exceeds decay. It is a typical thing seen in coniferous forests in cool climates.

Dysgenic

The term applies to the condition of being detrimental to the genetic quality of a population.

—xxx—

E

Easement

An interest or right granted by the owner to another party to the limited use of land.

Ecology

It is a branch of science that deals with the interaction of plants and animals with their environment.

Ecology

The study of interactions between organisms and their environment.

Ecosystem

Ecosystem is a system of interaction in living organisms (plant and/or animals), soil, and climatic factors where all are linked by the flow of energy and nutrients. An example of this interaction is the movement of nutrients from the soil through the roots into the tree; the manufacture of leaf material from some of these nutrients; the leaves dropping and decomposing; and the nutrients returning to the soil.

Ecotone

It is the transition area between two distinct communities which are present adjoining.

Edge

Edge is the boundary between two ecological communities, for example, field and woodland. Edges provide wildlife habitat. Consideration of an edge can reduce the impact of a timber harvest.

Elite Tree

A tree that has been shown by progeny testing to produce superior offspring.

Endangered Species

Refers to any species or subspecies which is in immediate danger of becoming extinct throughout all or a significant portion of its range.

Endemic

Generally it refers to a plant or animal species or subspecies native to a small region. Entomologically it is the normal population level of a destructive species.

Entomology and Forest

Entomology is the branch of science that deals with insects and their relation to forests and the products of the forest.

Environment

The surroundings or prevailing conditions that reflect the combined influence of climate, soil, topography, and biology (other plants and animals) present in an area which is extremely important in determining how well a particular species will grow in a given area is called environment.

Epidemic

A rapid widespread insect or disease incidence beyond normal proportions which is usually accompanied by excessive damage.

Epicormic Branching

They are branches that grow out of the main stem of a tree from buds which are produced under the bark. When epicormic branching is more, it increases knottiness and reduces lumber quality.

Epicormic Shoot

Epicormic shoot is a arises from a dormant or adventitious bud on the stem or branch of a woody plant.

Establishment

It refers to the process of developing a crop to the stage at which the young trees may be considered established, which ensures that the crop is safe from normal browsing and there is no need of special protection or special tending except cleaning, thinning, and pruning.

Eugenic

It refers to the condition of favorable to the genetic quality of a population.

Evapotranspiration

The process of loss of water in vapour form from a soil or open water surface (evaporation) or from surface of a plant through the stomata (transpiration).

Even-Aged

A type of forest where relatively small age differences exist between individual trees. The permitted differences in age is from 10 to 20 years. A larger differences up to 25% of the rotation age may be allowed for a stand that remains unharvested for 100 – 200 years old.

Even-Aged System

A silvicultural systems where the stands have an even-aged structure, e.g., clearcutting method, coppice method, seed-tree method.

Even-Flow Harvest

A harvesting scheme designed to extract exactly the same volume of wood fiber each period.

Evergreen Tree

A tree on which leaves do not die in the same year they appear.Evergreen, true cone bearing trees with needles, awl, or scale-like leaves, such as pine, spruces firs, and cedars belonging to the order coniferales and they are also referred to as "sol'twoods".

Ex situ Management

It refers to the management of planted stands of trees to protect them and insure survival, growth and identity; the conservation or preservation of trees as seed, pollen, tissue culture, or excised plant parts outside their natural habit.

Exotic

It denotes a non-native population introduced into a new area or a species that is introduced from another country.

Extension Forester

Cooperative Extension Service professional who educates woodland owners on how they can effectively manage their forests.

Extensive Forest Management

The protection of the forest from fire and insects; reliance on natural regeneration.

—xxx—

F

F_1

F_1 is the first (filial) generation of a cross which will be phenotypically uniform.

F_2

The second (filial) generation of a cross, produced by intercrossing or selfing the individuals from the F_1 generation. The progeny of an F_2 cross will be a segregating population.

Factor

A set or class of comparable treatments, for example, different amounts of nitrogen fertilizer, different methods of cultivation.

Factor Rating

It is a set of critical values which will indicate how well a land-use requirement is satisfied by a particular condition of the corresponding land quality.

Faller

A term synonymous to cutter, sawyer and chopper. A logger is a person who specializes in felling trees.

Fallow

A crop land that is allowed to lie idle, either tilled or untilled, during the whole or greater portion of a growing season where tillage is usually practised for controlling weeds and to encourage the storage of moisture in the soil. Land rested from deliberate cropping, not necessarily without cultivation or grazing but without sowing. It also includes the land left without a crop or weed growth for extended period, often to accumulate moisture.

Family

Family is a group of plants or animals that have some common characteristics. Eg. *Acacia* and *Albizia* trees are members of the same family, Mimosaceae. It is also the offspring of a single tree after open pollination, or of a single pair of trees after controlled pollination.

Family Selection

It refers to the selection of progeny families based on their mean performance. Here, the best individuals are usually selected in the best families.

Fan Design

It is a type of design for plant-spacing studies. In this design, the plants are arranged at systematically increasing distances along radii and arcs.

Farm Enterprise

It refers to an individual crop or animal production function within a farming system, the smallest unit for which analysis of resource use and cost return is normally carried out.

Farm Forestry

A system of growing trees for timber, poles, fuelwood on farmland is called farm forestry. It can be done in small **woodlots** or as boundary plantings.

Farm Household System

Refers to a system of a group of people, usually related, who individually or jointly provide management, labour, capital, land and other inputs for the production of crops and livestock, and who consume at least part of the farm produce.

Farm Services Agency (FSA)

The branch of the U.S. Department of Agriculture (USDA) that administers cost-sharing programs for such forestry practices as tree planting and timber stand improvement.

Farming System

It is a part of larger system, which is divided into subsystems. It is a unique and reasonably stable arrangement of farming enterprises that a household manages according to well-defined practices in response to the physical, biological and socioeconomic environments, and in accordance with the household's goals, preferences and resources. The different factors combine to influence output and production methods. It also refers to a decision-making unit which comprises the farm household and cropping and livestock systems, that produces crop and animal products for consumption and sale. The farming system is a subsystem of a higher level land-use system such as village or watershed, which in turn forms a component of the agricultural sector of the regional system.

Farming Systems Research and Development

The term is applied to an approach to agricultural research and development that views the whole farm as a system and focuses on the interdependence among the components under the control of the farm household's members and how these various components interact with the physical, biological and socioeconomic factors not under household control. It also involves selecting targets, areas and farmers, identifying problems and opportunities, designing and executing on-farm research, evaluating results, and extension.

Farmyard Manure

It is a type of manure which includes the partly decomposed excreta of domestic animals mixed with straw or other litter.

Fast-Growing Tree

It refers to a tree species that matures quickly and is usually not long lived. These trees often be highly productive on fertile sites and they can yield 15–20 t ha^{-1} of wood products per annum or more. Eg. leguminous species in the Mimosaceae and Papilionaceae families, actinorrhizal plants such as the genus *Alnus,* which are nitogen fixing and *Calliandra, Gliricidia, Leucaena* and *Sesbania.*

Feathering

It is a process of stimulation of the production of lateral branches in nursery stock.

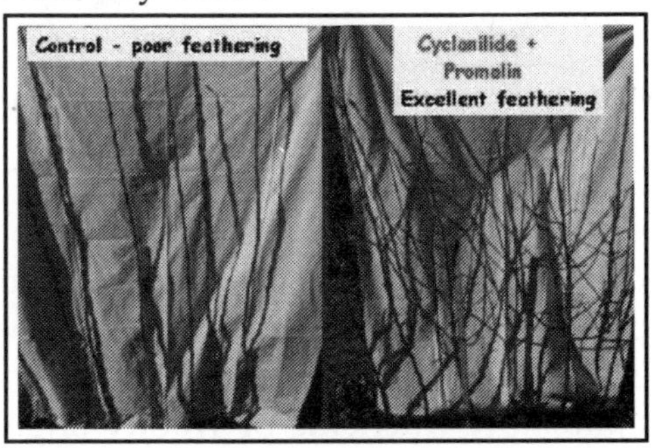

Feed Lot

A plot of land where beef cattle are confined and fattened by feeding high-protein feeds, for example, silage.

Fertilising

The practice of the application of chemical or organic fertilisers done to increase the unit area soil productivity.

Fertilization

It refers to the union of male and female gametes to produce a fertilized egg cell in floral biology but in agriculture it refers to the practice of adding nutrients to soil or plants for use by plants.

Summary of Fertilization

1. A pollen tube grows from pollen grain, down the style, through the micropyle and into the ovule.
2. Two sperm nuclei are discharged into the embryo sac resulting in double fertilization.
3. The fertilized egg (zygote) will develop into the embryo.
4. The ferilized central cell will develop into endosperm.
5. The ovule develops into a seed.
6. The ovary develops into a fruit containing one or more seeds.

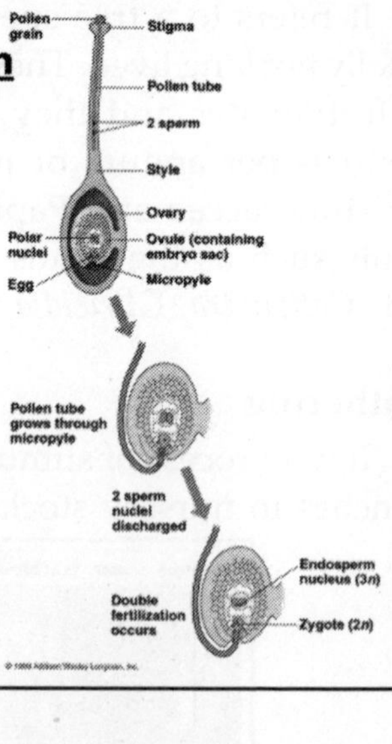

Fibre

It is a general term of convenience for any long, narrow cell of wood other than vessel elements and parenchyma. Includes the tracheids of gymnosperms and the libriform wood fibres and fibre tracheids of woody angiosperms.

Field Capacity

The term refers to the amount of water a soil will hold against gravity when given reasonable time to drain. Commonly taken as one-third atmosphere tension for clays, and one-tenth atmosphere for sands. In other words, it is the total amount of water remaining in a freely drained soil after the excess has flowed into the underlying unsaturated soil unit.

Field Layer

The vegetation at ground level in a forest or woodland.

Field Nursery

A loose term for a nursery, generally not permanent, established in or near the forest rather than near an administrative or executive headquarters. Also referred to in Ontario as **satellite nursery**.

Fill Planting

The term applies to the planting of trees in areas of inadequate stocking to achieve the desired level of stocking, either in plantations or areas of natural regeneration.

Filter Strip

A strip of permanent vegetation of sufficient width and vegetative density above farm ponds, diversion terraces and other structures to retard flow of runoff water, causing it to deposit soil, thereby preventing silting of a storage structure or reservoir.

Final Crop

Final crop is the trees that remain after successive thinnings and that are finally felled at maturity.

Final Cutting

It applies to the last cut practiced in a series of progressive regeneration which removes the last of the original seed trees when the regeneration is considered established.

Fire

Fire is used in forestry and range management to carry out a **burn** in a controlled manner and to dispose off unwanted vegetation. A headfire is where the fire is lit at the bottom of a slope and allowed to burn quickly to the top. A backfire is lit at the top and burns slowly downhill. Flankfires are lit at the sides; jackpot fires are where only piled **slash** is burned. A wildfire is a burn occurring spontaneously (that is an uncontrolled fire).

Firebreak

Fire break is a form of fire protection, which is done to prevent uncontrolled fires, where either a zone of trees and any other woody vegetation is removed so that only a limited amount of flammable material is present. Moreover tree species that are relatively less flammable are planted (for example, broadleaved) so as to separate more flammable species (for example, resinous conifers).

Fire control

It refers to all activities concerned with the suppression of a forest fire.

Fire Danger

Fire danger is the result of both constant and variable factors that determine whether fires will start, spread, and do damage, and the estimated difficulty to control.

Fire Hazard Reduction

It refers to all the activities concerned with the treatment of fuels that reduces the threat of ignition and spread of fire.

Fire Line

A type of fire break where a trail around a fire is dug down to mineral soil and all debris are cleared.

Fire Pack

It is a pack used by firefighters on their backs to carry tools, equipment, and supplies.

Fire Prevention

It is the control of fire through education, hazard reduction, and law enforcement to reduce the number of fires.

Fire Resistant Clothing

It refers to the trademark yellow shirts and green pants of wildland firefighters.

Fire Scar

It is a injury or wound in the bole of a tree caused or accentuated by fire.

Fire Season

The period or periods of the year during which fires are likely to occur, spread, do sufficient damage, and otherwise warrant organized fire control.

Fire Shelter

Fire shelter is used by the firefighters for personal protection as a last resort if a wildfire traps them and they cannot escape. Firefighters can get into the tent-like shelter, made of heat reflective material, in about 25 seconds.

Fire Shovel

Fire shovel is a tool specifically designed for constructing a fireline, feature a tapered blade with both edges sharpened for scraping, digging, grubbing, cutting, and throwing dirt.

Fire Suppression

It is the activity of extinguishing a fire after its detection. In direct method where the edge of the fire is extinguished directly, whereas in indirect method where the control line is located along a favorable firebreak and the intervening strip between the fire and the firebreak is backfired. One-lick method is a system of managing personnel on a fire, where the entire crew constructing the control line moves forward without changing relative positions in the line. As they move forward, they do "one lick of work," then advance one or more steps. The number of steps is controlled primarily by the number engaged and the consequent proper spacing of licks, in order that the control line may be completed and the fire extinguished when the last person has passed over the line.

Fire Weather Index (FWI)

FWI is the numerical rating of fire intensity that combines the expected rate of spread and the total amount of available fuel for combustion which is suitable as a general index of fire danger.

Fitness

Fitness is the relative ability of organisms of a particular genotype to survive and produce offspring, or the contribution of one of a pair of alleles to the general vigour of an organism. In other words, it is the degree to which the characteristics of any particular plant genotype confers the ability to establish and maintain the plant in a particular ecological niche.

Fitness and Adaptive

It is defined as the ability of a population to survive, reproduce, and adapt genetically to a changing environment for at least a modest number of generations.

Fitness and Evolutionary

It is defined as the ability of a population to survive, reproduce, and adapt genetically to a changing environment for many (usually more than 50) generations.

Fixation

Fixation is a process in a soil by which certain chemical elements essential for plant growth are converted from a soluble or exchangeable form to a much less soluble, or to a non-exchangeable form and made available.

Flash Point

The temperature at which a material will burst into flame is called as flash point.

Flexibility

The term is defined as the degree to which a plant genotype confers the ability for the plant to establish and maintain itself in more than one ecological niche.

Flow Chart

It is a diagrammatic representation a series of steps or procedures in logical sequence to achieve a stated objective or objectives with conventional symbols, of the structure of a system in terms of physical and information flows between compartments.

Flower

Flower is a reproductive structure which bears pistils or stamens or both, and usually sepals and petals.

Flower Induction

It is the process of stimulation of plants to produce flowers.

Flushing

Flushing is the sudden occurrence of the obvious growth (stem elongation, bud burst and leaf expansion) of leafy shoots.

Fodder

The Plants or plant parts eaten by browsing or grazing animals which includes leaves, flowers and pods. Eg. *Acacia, Leucaena, Prosopis.*

Foliage

It refers to the entire leaf mass of a tree or trees (or of plants generally).

Forage Forestry

Growing of close-planted fast-growing tree species which can be harvested mechanically by a forage harvester for wood products such as woodchips.

Forage

It commonly refers to grasses and legumes.

Forage Value

It is the relative importance for grazing purposes of a range plant or plants as a whole on a range.

Forb

Forb is a non-grass broadleaved herb.

Forest

Forest is defined as a continuous stand of trees above 10 m high, with interlocking crowns. Forests can be classified into many types on the basis of ecozone, vegetation type, climate, dominant species and conformation.

Forest Fragmentation

It refers to the subdivision of large natural landscapes into smaller, more isolated fragments. Fragmentation affects the viability of wildlife populations and ecosystems.

Forest Garden

It refers to a land-use form on private lands outside the village where trees are planted and sometimes additional perennial crops occur.

Forest Genetics

It is defined as the study of heredity and variation in forest trees.

Forest Grazing

An integrated farming system **(silvopastoral)** where timber-producing trees and grazed pasture are grown together with a prime objective to increase long-term net profit per hectare. Growing pasture under trees as an alternative source of income to production thinning of timber or pulpwood.

Forest Management License Agreement

It is the contractual agreement between the Provincial Government and a lumber or pulp company to provide a wood supply to the processing plants of that company which ensures that the forest on such lands are harvested and reforested to produce successive crops of timber on a sustainable-yield basis.

Forest Management

The management of forest property by the application of business methods and technical forest principles.

Forest Management Unit

It applies to an area of forest land managed as a unit which can be the entire province or territory, a provincial forest management subdivision, an industrial timber limit for fibre production and other renewable resources.

Forest Model

It is a computer-based simulation that within definable parameters forecasts the development of the forest.

Forest Nursery

It is defined as an area in which young trees are grown for forest planting.

Forest Practices Act

It refers to the legislation regulating private forest harvest to reasonably for assuring adequate regeneration and protection of soil and water values.

Forest Product

It refers to poles, roundwood, medicinals, gums, resins, oils, fungi and honey that are obtained for use from a forest.

Forest Protection

The protection activities connected with the prevention and control of damage to forests from fire, insects, disease, and other injurious and destructive sources.

Forest Site

It is a land unit characterized by different factors like climate, soil, and topography that control forest type and growth of a tree.

Forest Survey

It refers to the measuring and marking property of a forest boundaries and the inventory of forest land to determine acreage, condition, timber volume, and species, for specific purposes (such as timber purchase and forest management) or as a basis for forest policies and programs. Also refers to carefully measuring and marking property boundaries.

Forest Type

A descriptive term used to group stands of similar character in composition and development, to differentiate them from other groups of stands.

Forest Utilization

A branch of forestry which is concerned with the operation of harvesting, processing, and marketing the forest crop and other forest resources.

Forest Tree Breeding

It is defined as the genetic study of trees to solve some specific problem or to produce a specially desired product.

Forest Tree Improvement

It is the control of parentage combined with other silvicultural activities (such as site preparation or fertilising) to improve the overall yield and quality of products from forest lands.

Forestation

It is defined as the establishment of forest naturally or artifically on areas where it has been absent or insufficient.

Forester

A person who has been professionally educated in forestry at a college or university.

Forested Wetland

It is an area characterized by woody vegetation taller than 20 feet where soil is at least periodically saturated or covered by water.

Forestry

Forestry is defined as a practice embracing the science, business, and art of creating, conserving, and managing forests and forest lands for continuous use.

Forestry Incentives Program (F.I.P.)

A Federal cost-sharing program administered by the NRCS that reimburse part of the costs landowners incur in completing certain forestry practices.

Form Class

It is a measure of bole taper derived by dividing diameter inside bark at a given height (usually 16 or 32 ft) by d.b.h. These values are often required to use tree-volume tables.

Form Pruning

The process of early pruning of trees being grown for timber to remove unwanted (multiple) stems, leaving the straightest and most vigorous.

Form

The shape of a log or tree. Trees with good form are straight, more or less cylindrical, with fine branches, and do not taper rapidly.

Formulation

It is defined as the process by which herbicidal compounds or insecticidal sprays are prepared for practical use with dispersants, 'stickers' etc.

Forward Selection

It is the selection or choosing the best individuals out of a progeny test for use in seed orchards and/or subsequent generations of breeding.

Foundation Seed

Foundation seed is the seed stock produced from breeder's seed by or under the direct control of an agricultural experiment station. Foundation seed is the source of **certified seed,** either directly or through **registered seed.**

Fractal

An object (actual or mathematical) in which form is independent from scale, that is, which always repeats the general pattern. Plant canopies are considered fractal.

Free Growth

Free growth is the situation in which a tree or other plant has grown with its crown more or less free from competition. It should also include freedom from below-ground competition as well.

Free Living

The term refers to a microorganism that lives freely unattached, or a pathogen living in the soil outside its host.

Free Thinning

Removal of trees to benefit best trees, regardless of crown class.

Free-To-Grow

A term synonymous to free growing which refers to the condition of a forest stand when it is established and acceptable for entry into the productive timber land base. The stand should meet the criteria like minimum stocking, desired species composition, minimum height development, and freedom from competition that impedes growth.

Friable

The term pertains to soils that, when either wet or dry, crumble easily between the fingers.

Frill Girdling

It refers to the practice of Girdling by making a series of downward, more or less overlapping incisions, generally for the introduction of poison. Spaced incisions are termed frill cuts. A double series of such incisions is referred to as double-frill girdling.

Frill

It is the V shaped cut in the cambial tissue of the tree made with a machete or other sharp tool which used as a place to apply herbicides.

Frilling

The method of killing trees by inflicting a series of cuts around the bole and applying an herbicide to the wounds. Frilling or girdling can be used to reduce the density of a stand or to kill individual undesirable trees.

Frost Crack

Forest crack is longitudinal crack on the outside of a tree, caused by extreme cold which is especially common on thinbarked species, such as hemlock and true fir.

Frost Pocket

It refers to the depression in the terrain into which cold air drains but cannot escape which exposes the area to freeze injury.

Fruit Set

The term applies to the physiological process in which a fertilized ovule becomes 'committed' to further development.

Full-Sibs

Full sibs are the trees which have both parents in common.

Fungicide

Fungicides are agents which are used to kill or inhibit the growth of fungi and their spores.

Fusee

Fusee are colored flares used by the firefighters to ignite fires in burnout operations.

—xxx—

G

Gall

Gall refers to a pronounced localized swelling of greatly modified structure which occurs on plants as a result of irritation by a disease or insect.

Gallery

It is a passage or burrow, excavated by an insect under bark or in wood for various purposes like feeding or egg laying.

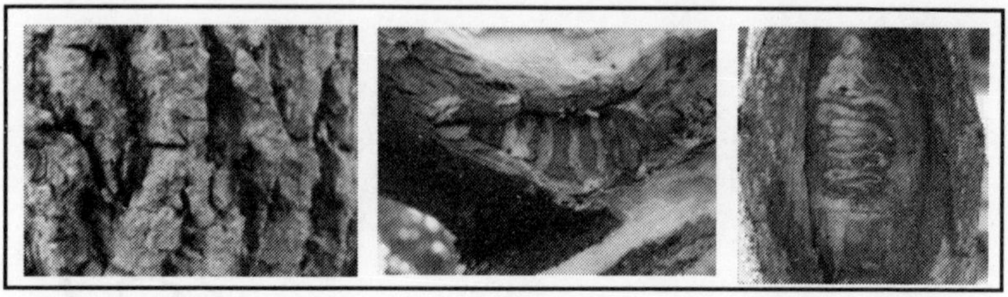

Gallery Forest

It is the term synonymous to riparian forest which applies to the vegetation, with trees and shrubs, growing alongside or close to a watercourse, lake, swamp, or the like, and often dependent on its roots reaching the watertable.

Gamete

Gamete is a haploid male or female reproductive cell, produced through meiosis and capable of uniting in the process of fertilization.

Gene Bank

The place for the maintenance of germplasm in the form of seeds, tissues or as growing plants with the appropriate facilities and trained staff.

Gene Complex

Refers to the balanced system of genes constituting the internal environment within which each gene acts, so balanced that a change in one affects the operation of other genes.

Gene Fixation

It is the condition in which a particular allele becomes the only allele present in a population, because of selection or genetic drift.

Gene Flow

Gene flow is the exchange of genetic traits between populations by movement of individuals, gametes or spores or the spread of genes through crossing.

Gene Frequency

It is defined as the proportion in which alleles of a gene occur in a population.

Gene Pool

It is defined as the total genetic information possessed by the reproductive members of a population of sexually reproducing organisms.

Gene

The unit of inheritance.

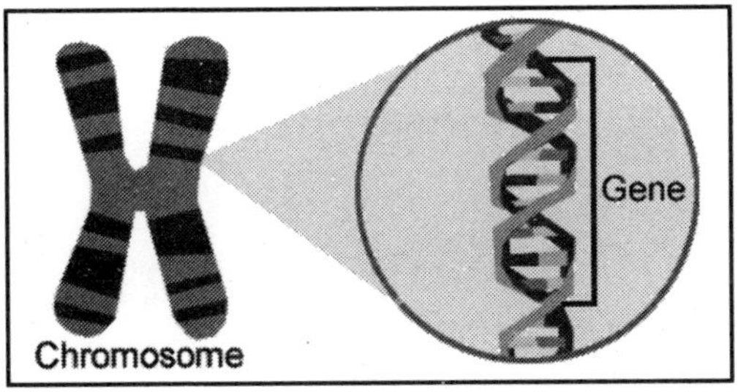

Genecology

Genecology is the combination of ecology and genetics to study the genetic variation among populations of a species that is correlated with habitat.

General Combining Ability

General combining ability (GCA) is the ability of an individual to produce progeny with high genetic quality, when crossed to many other individuals in the population.

Genet

The ramets derived from one seed.

Genetic Base

Genetic base is the aggregate of genetic variability available within a cultivar or species.

Genetic Drift

It is defined as the change in gene frequency and population characteristics caused by chance rather than selection in small populations. Random fluctuations in gene frequency caused by sampling error.

Genetic Engineering

Genetic engineering is the alteration of the genetic composition of a cell by various procedures like transformation, protoplast fusion in tissue culture.

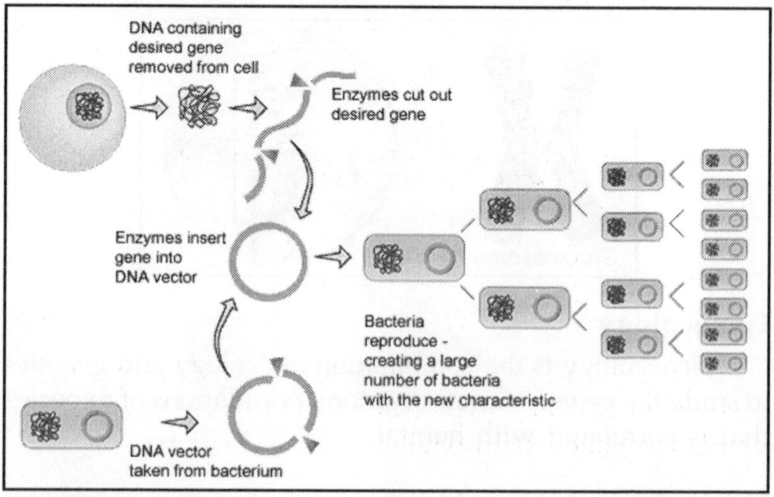

Genetic Gain

The increase in productivity following a change in gene frequency is called as genetic gain.

Genetic Plasticity

It is defined as the capacity for adaptation caused by genetic changes; as in natural selection.

Genetics

The branch of biology which deals with the causes of resemblances and differences among organisms related by descent.

Genome

It refers to one complete set of genes containing all the genetic information to produce an individual.

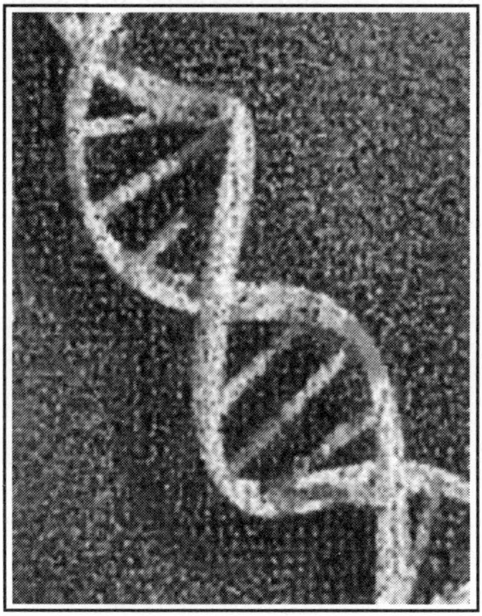

Genotype

Genotype is the entire genetic constitution of an organism which is determined chiefly from the performance of progeny and other relatives.

Genotype x environment = phenotype.

Genotype x Environment Interaction

It is defined as the degree to which relative ranks (levels of differences) are found when sets of genotypes are tested in different environments (or cultural conditions). In other words it is the tendency for the size of a difference in performance between two genotypes to depend on environment.

Genus

Genus is a rather arbitrary category in the taxonomic hierarchy between that of family and species which includes one or more closely related species, plant genera are defined mostly on characteristics of the flower, fruit or both.

Geographic Information System (GIS)

It is defined as an information system that uses a spatial database to provide answers to queries of a geographical nature through a variety of manipulations, such as sorting, selective retrieval, calculation, spatial analysis, and modelling.

Geographic Variety

It is a subdivision of a species with distinct morphological characters and a distinctive geographical range, given a Latin name according to the strict rules of the International Code of Botanical Nomenclature. A taxonomic variety is known by the first validly published name applied to it and that nomenclature tends to be stable.

Geophyte

Geophyte is a herbaceous plant, perennial or biennial, that periodically reduces its shoots to storage organs embedded in the ground. This often occurs during the season that is unfavourable to the whole shoot system.

Geotropism

Geotropism is the movement of plant growth induced by gravitational stimulus.

Germination

Germination is the process of the growth of the embryo in the seed until the emergence of the embryonic radicle through the seed coat. The capacity of the embryo to emerge from the seed coat with the essential structures indicates a potential to produce normal plants. Germination follows imbibition (absorbing water and swelling) in case of dry seeds.

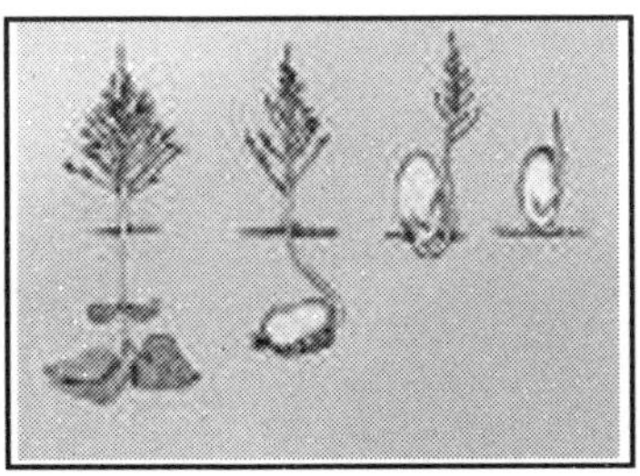

Germinative Capacity

It is the percentage of seeds that germinate during a specified time interval, determined by the maximum rate of germination.

Germplasm

Germplasm is defined as the sum total of the genes and cytoplasmic factors governing inheritance. The hereditary material transmitted to offspring through the germ cells. It includes the materials like seeds, cuttings and tissue cultures.

Girdle

It is the practice of encircling the stem of a living tree with cuts that completely sever bark and cambium as well into the outer sapwood, it is done to kill the tree by preventing the passage of carbohydrates to the roots.

Girth

Girth is the size or measurement around an object, such as the circumference of a tree trunk.

Gley

It applies to soil with impeded drainage. Used in soil classifications based on profile leaching.

Global Radiation

Global radiation is the total of direct solar radiation and diffuse sky radiation received by a unit horizontal surface.

Grade

Grade is a system of classifying lumber or logs according to quality. It also applies to the steepness of a forest road.

Graded Terrace

The term refers to a terrace having a constant or variable grade (slope) along its length.

Gradoni Bench

Small bench terrace or narrow shelf cut along the contour which is usually cut with an inward slope is called gradoni bench.

Graft Incompatibility

It is the failure of the graft union between the stock and the scion.

Grafting

A method of plant propagation by transplantation of a bud or a scion of a plant onto another plant, the joining of cut surfaces of two plants, placing a portion of one plant in close cambial contact with another to obtain vegetative union between the two plants.

Grain

Grain refers to the direction, size, arrangement, appearance, or quality of the fibers in wood.

Grassland

It refers to the land covered with grasses and other herbaceous species. Here woody plants may be present, but if so,they do not cover more than 10% of the ground. Grassland are classified based on ecozone, topography, climate, soil conditions, and so on. Derived grassland is maintained in that condition by regular burning; edaphic grassland arises on particular soil types, for example, those found in or around permanent or seasonal swamps.

Gravelly

Gravelly refers to the soil texture where soil mass contains 15–90% pebbles by volume.

Gravimetric Method

It is a density determination (eg of wood) by weight and volume, as opposed to densitometric determination by beta-ray or x-ray analysis.

Grazing

Grazing is a method of feeding by herbivores characterized by repeated removal of only a part, generally the leaf, of the plant, which is most commonly called herbage. Defoliation by herbivorous animals.

Grazing Capacity

The ability of a range unit, in years of normal rainfall, to give adequate support to a constant number of livestock for a stated period each year without deteriorating. It is expressed in number of livestock per acre of given kind or kinds, or in number of acres per specified animals.

Green Lumber

The lumber with the moisture content greater than that of air-dried lumber and unseasoned lumber, boards from logs processed through mill before drying is called green lumber.

Greenhouse Effect

It is the steady warming of the earth's atmosphere, caused by the accumulation of greenhouse gases, including carbon dioxide, methane, carbon monoxide etc.

Green Manure

Green manure is a crop that is grown for soil protection, biological nitrogen reduction, or organic matter and ploughed, disked or hoed into the soil for the improvement of the soil.

Gregarious

The term which applies to species that occur in groups within a community and that may form pure stands of species in the natural state.

Gross Energy Value

It is the energy released when a known weight of food is oxidized in a bomb calorimeter, that is, the total heat produced in animal feed. The energy value of different components are proteins - 24 MJ kg^{-1}, fats -38 MJ kg^{-1} and carbohydrates - 17.5 MJ kg^{-1}.

Gross Plot

Gross plot is the full plot size, including guard rows.

Ground Cover

Ground cover is a crop planted to provide a covering over the soil, in other words it applies to any vegetation producing a protective mat on or just above the soil surface. In forestry, low-growing shrubs and herbaceous plants under the trees are called as ground cover.

Ground Flora

The term includes the grasses, herbs and other plants growing on the forest floor.

Ground Truth

Ground truth is the data collected by field observation to compare with, or support, the interpretation of remote sensing imagery.

Ground Validation

It is the confirmation of information from satellite imagery or aerial photography.

Ground Water

The water below the land surface that has drained through the upper soil layers is called as ground water.

Group Selection

A process of harvesting patches of trees to open the forest canopy and encourage the reproduction of uneven-aged stands.

Group-Selection Method

It refers to a method of regenerating uneven-aged stands in which trees are removed in small groups.

Grove

A small wood or group of trees is called groove.

Growing On

Growing on is the term used for the re-establishment of seedlings or young plants into a more spacious site or container than that in which they were growing previously.

Growing Season

Growing season refers to the period(s) of the year during which water is available and temperatures are favourable (not too high or low) for the plant growth.

Growing Stock

Growing stock applies to all the trees growing in a forest or in a specified part of it. It is expressed in terms of number or volume.

Growth Correlation

Growth correlation is the interrelationships between the various growth processes that control the rate, extent and kind of growth that occurs in an organ or plant.

Growth Habit

The basic pattern of structural development or symmetry attained by a plant, for example 'shrubby', 'treelike' or 'viney', 'erect' or 'spreading' is called growth habit.

Growth Rate

Growth rate is defined as the rate at which wood has been added to the tree at any particular point. It is expressed in the number of annual rings per inch.

Gyppo Logger

The term synonymous to contract logger used for a self-employed, independent timber harvesting contractor who is not an employee of the log purchaser.

Growth Regulator

A natural substance that regulates the plant growth by enlargement, division or activation of plant cells. Gibberellin, cytokinin, auxin etc; are examples.

Growth Retardant

Growth retardants are chemical substance used to slow down the growth of crop plants. Eg. Maleic hydrazide, ABA, CCC, dormin.

Growth Rings

Growth ring refers to the layers of wood a tree adds each season which are frequently visible when a tree is cut and can be used to estimate its age and growth rate.

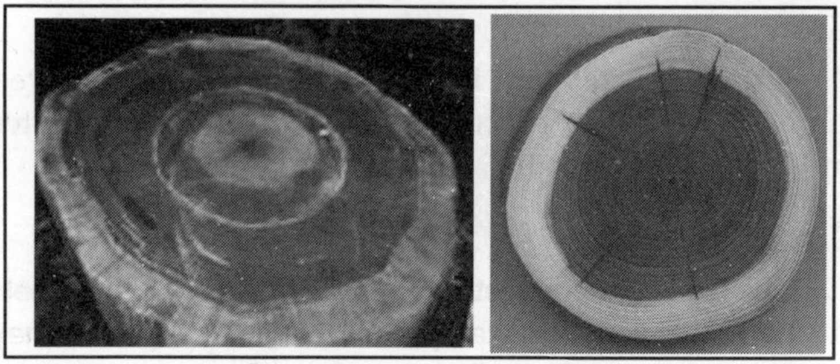

Grubbing Out

The process of felling trees and shrubs by exposing and cutting the roots. It is also called as grub felling.

Guard Row

Guard row is a line of plants along the edge of a research plot that is not measured and is used to ensure that the measured part of the plot (net plot) is not affected by external influences.

Gully

It is the channel cut by concentrated runoff but through which water commonly flows only during or immediately after heavy rains, or during the melting of snow. It is sufficiently deep not to be obliterated by normal tillage operations.

Gully Erosion

The erosion process in which water accumulation occurs in narrow channels over short periods and removes the soil from this narrow area, sometimes even to a considerable depths.

Gum

It is a complex polysaccharidal substances which is formed by cells in reaction to external wounding or infection in the cells.

Gummosis

The term applies to the production of gum by or in plant tissue which maybe sometimes caused by bacterial infection.

Guttation

Guttation is a process by which plants exude liquid water from the leaves in excess of transpiration through the stomatal opening.

—xxx—

H

Habitat

Habit is the natural environment of an organism excluding other living organisms (the biotope) or the place or the environment in which the plant or animal lives is called habitat.

Half Sib

The tree with one parent in common with another (usually the female).

Halophyte

A plant that is tolerant of high levels of mineral salts in the soil, including sodium chloride and it can withstand salt spray if near the sea is called halophyte.

Hand Planting

Hand planting is a reforestation method of planting seedlings by hand, usually with spacing for minimize competition and maximizing growth. Seedlings are often 2 years old, giving the new forest a head start over seeding methods.

Haploid

It denotes a cell or organism with the gametic chromosome number (n).

Haptotropism

It is the movement or growth of a plant organ as a response to contact or a mechanical stimulus.

Hardboard (Fibreboard)

The sheet of material commonly with 0.25 cm thickness made from pulped wood is called hardboard.

Hard Seed

The seed with thick and tough testa, which delays water penetration and germination is called as hard seed.

Hardening Off

It is the process by which young seedlings are prepared for transplanting by gradually reducing water, shade or both, and thus causing changes in the leading shoot that renders it more resistant to drought and cold. It also refers to the treatment of tender plants to enable them to survive a more adverse environment by withholding nutrients, lowering temperatures, allowing temporary wilting, and other methods to slow growth rate.

Hardiness

The term applies to the attribute of a plant that enables to withstand environmental stresses (drought, frost, exposure to heat, and so on).

Hardpan

A cemented (indurated) or hardened soil horizon. This horizon, which may have any texture, is compacted or cemented by iron oxide, silica, sesquioxides, calcium carbonate and organic matter or another substance. Hardness or rigidity is maintained when wet or dry, and samples do not disintegrate in water.

Harvest

Harvest is the process of extraction of some type of product from the forest which is generally associated with a cutting.

Harvest Age

It is the duration or the number of years taken by a tree from establishment to maturity.

Harvest Increment

It is the increment in the harvested part, or the above-ground biomass, of a woody perennial plant over a stated period.

Harvest Index

Harvest index is the proportion of the total plant biomass (or more usually the above-ground biomass) produced by a crop species that forms a particular, required (harvestable) plant part; it is expressed as a percentage.

Hardwoods

It is a general term encompassing broadleaf, deciduous trees.

Haulback line

The line used to pull chokers or the carriage from the landing out to the felling area in cable logging is called haulback line.

Heading Back

The process of removing the apical part of a branch (stem), and so its apical dominance to stimulate the growth of lateral branches during pruning.

Headlamp

Headlamp is used by the firefighters who work on the fireline at night. They wear these flashlights on the front of their helmets.

Headland

A strip of unploughed land at the end of a set of furrows or along a fence is called headland.

Heart Rot

A decay characteristically confined to the heartwood. It usually originates in the living tree.

Heartwood

The wood in the centre of a tree, no longer in use for conducting water from the roots to the leaves which is often darker in colour than the outer wood (sapwood) and may contain chemicals that make it more resistant to decay is called as heartwood.

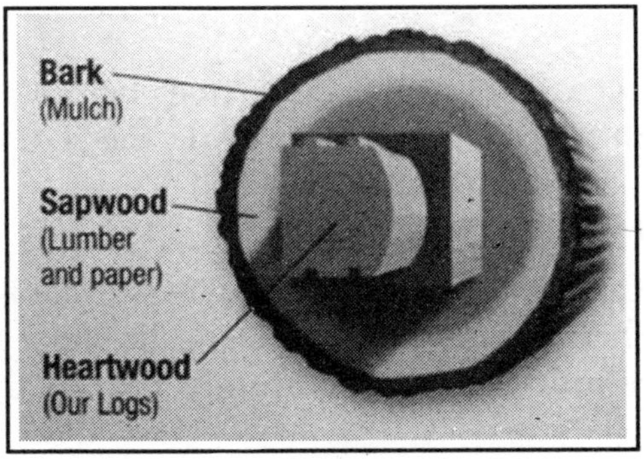

Hedge

Hedge refers to the bushes or shrubs or trees planted in a row and trimmed which is used to separate one piece of land from another.

Hedge row

A barrier of bushes, shrubs or small trees growing close together in a line.

Heel-in

The process of storing young trees before planting by placing them in a trench and covering the roots with soil.

Hemicryptophyte

It is defined as a self-supporting plant that is perennial or biennial, herbaceous and with periodic shoot reduced to a remnant shoot that lies relatively flat on the ground surface.

Hemi epiphyte

The term refers to a plant that grows by supporting itself on others and that roots in the ground, germinating on other plants first, or that first germinates in the ground, grows up a tree and then later disconnects its soil contact. Eg. Pseudo-liana.

Herbaceous Layer

The layer or part of the understorey in a forest that consists mainly of non-woody plants is called herbaceous layer.

Herbaceous Perennial

It is a plant with soft, succulent stems whose top is killed back by frost in many temperate and colder climates but whose roots and crowns remain alive and develop new top growth when favourable growing conditions return.

Herbage

Refers to the vegetative parts of plants.

Herbicide

It is a chemical substance that can kill or suppress the growth of certain plants; a weedkiller.

Herbivore

Animal that feeds on plant material including insects is called herbivore.

Heritability

Heritability in a broad sense that portion of the total variance caused by genetic factors. In a narrow sense, that portion of the total variance caused by genes with additive effects and most indicative of the superiority that can be transmitted by seed.

Heterophyllous

The condition in presence of leaves of different shapes occurs on the same plant.

Heterosis

The phenomenon in which the hybrid vigour is exhibited when the mean F_1 hybrid phenotype falls outside the range of the parents. It means an increase in some measure over the parents in statistical terms.

Heterozygous

The condition of possessing different alleles at a particular locus on the chromosome which is derived from the union of gametes of dissimilar genotype. A heterozygous individual is called a heterozygote.

Heuristic

It is an approach to decision making, or investigation, whereby the operator reaches a conclusion with reference to past experience, or inference, sensory, contextual or coherent information.

Hierarchical Classification

It refers to the classification in which the classes are ranked and every individual belongs to a class and every class to a higher ranking class, up to the highest class, which is the totality of all individuals.

Hierarchy

Hierarchy is the structural relationship in which each unit consists of two or more subunits, the latter being similarly subdivided. It also denotes a model of an agricultural system involving units (systems or subsystems) arranged according to increasing scale and complexity, ranging from the plant cell at the lowest and the region or nation at the highest levels.

High Forest

It applies to the crops and stands of trees, generally of seedling origin, that normally develop a high closed canopy. A term originally used to differentiate the natural, essentially seedling forest of long rotation from the artificial, coppice forest of shorter rotation.

High Forest System

It is a silvicultural system which produces forest trees from seedlings rather than from coppice.

High Grading

The removal from the stand of only the best trees, often resulting in a poor-quality residual stand.

High Density Planting

The term applies to planting of trees or crops in a close manner.

High lead Logging

It is a logging system that uses cables rigged to a spar high above the ground so that one end of the logs can be lifted during yarding.

Hinge Wood

The portion of the tree that remains uncut during felling is called hinge wood. The width and location of this wood helps determine which way the tree will fall.

Hog

Hog is a machine used to reduce waste pieces of lumber and slabs, or small tree stems, to chip form.

Hill Farming

The term refers to the system of farming in broken terrain where fields are often small and the land is sloping. Soil erosion is a major problem in hill farming.

Home garden

It is a land-use form on private lands surrounding individual houses with a definite fence, where several tree species are cultivated together with annual and perennial crops with the inclusion of small livestock. There are many forms of such gardens such as 'village forest gardens', 'compound gardens', 'kitchen gardens'.

Homeostasis

The phenomenon of maintenance of static or dynamic stability irrespective of external effects is called homeostasis.

Homozygous

The condition of occurrence of alleles like at corresponding loci on homologous (corresponding) chromosomes derived from the union of similar gametes. An individual in homozygous condition is called homozygote.

Horizontal Resistance

It refers to multigenic, partial resistance equally effective against all races of a pathogen. It is also called as field resistance.

Hot Deck

Hot deck is a log pile where both yarding and haul-truck loading take place in rapid succession.

Humid Tropics

Humid tropics are the tropical areas with excessive moisture. However, excessive moisture generally occurs only during a certain period of the year, while drought may occur in the same area during another part of the year. The humid tropics are characterized by isothermal conditions, which implies that temperatures in these areas are generally suitable for year-round crop production. Total radiation varies from relatively low values during the wet period to relatively high values in the dry season.

Humification

It is the process involved in the decomposition of organic matter which leads to the formation of humus.

Humus

The organic matter, including lignin, that has reached a more or less stable, advanced stage of decomposition which is usually characterized by its dark colour, considerable content of nitrogen, a carbon-to-nitrogen ratio approaching 10:1, and by various physical and chemical properties, such as high base-exchange capacity, water absorption and swelling. It also refers to the lower part of the litter layer, consisting principally of amorphous organic matter and located immediately above the A horizon.

Hybrid

Hybrid is the progeny or individual derived from seed obtained from natural or artificial fertilization, between plants belonging to different forms, species or genera. In other words it applies to the first generation cross between two genetically diverse parents.

Hydraulic Conductivity

Hydraulic conductivity is an expression of the readiness with which a liquid such as water flows through a solid such as soil in response to a given potential gradient.

Hydrology/Engineering

It is the process of removal of water from soil, particularly by surface run-off and subsurface percolation and artificially by measures for hastening removal, e.g., by ditching.

Hydromorphic Soil

It refers to the soils developed in the presence of excess water.

Hydrophyte

The plant that grows in excess water or wet soil is called hydrophyte.

Hygrometer

It is an instrument for measuring the relative amount of moisture in the atmosphere and for the determination of dewpoint.

Hypocotyl

It is the part of the stem bearing the cotyledons. The germination is hypogeal if both the cotyledon and the hypocotyl remain buried in the soil.

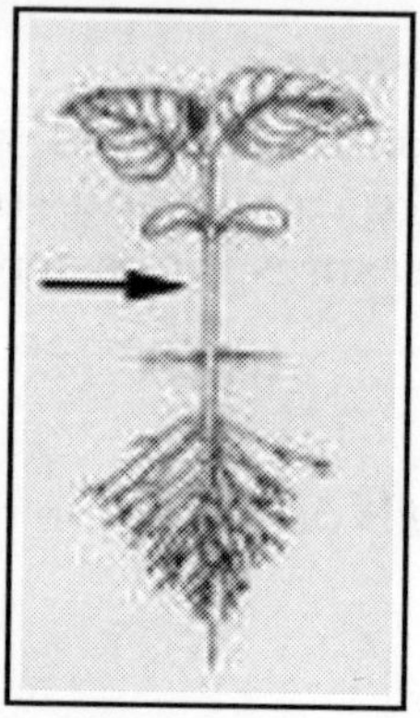

Hypsometer

An instrument used for measuring the heights of trees by employing geometric or trigonometric principles is called hypsometer.

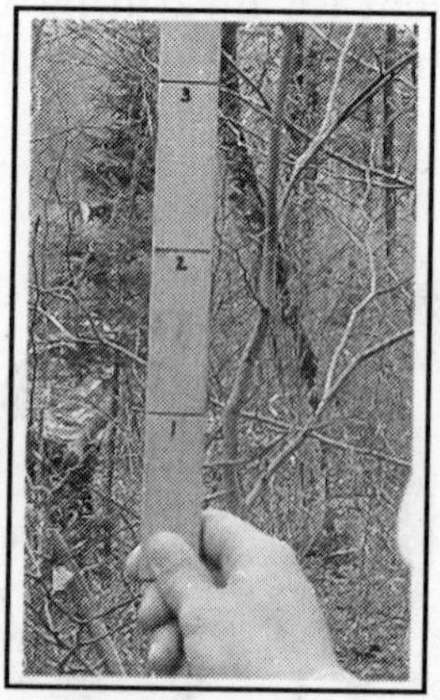

—xxx—

I

Ice Damage

The damage by an ice storm which causes breakage of tops and branches and stripping of branches and needles.

Ideotype

Ideotype is a conceptual model of a plant type that will be best suited to a particular set of circumstances. Ideotypes can be defined in terms of both form and function. There can be 'isolation', 'competition' and 'crop' ideotypes.

Ignite

It is a process to set fire or a cause to burn.

Illuvial Horizon

It is the soil horizon which receives material in solution or suspension from some other part of the soil.

Imbibition

Imbibition is the process in which the seeds absorb water from their surroundings or the stage at which they do it, for example, after a dry seed has been planted in moist soil.

Immature

Refers to those trees or stands that have grown past the regeneration stage but are not yet mature is called immature.

Immobilization

Immobilization is the conversion of an element from the inorganic to the organic form in microbial tissues, thus rendering the element not readily available to plants or other microorganisms.

Impeded Drainage

The term refers to the restriction of the downward gravitational movement of water which result in the development of anaerobic conditions due to waterlogging.

Impervious Soil

Impervious soil is the soil through which water, air or roots cannot penetrate. No soil is even completely impervious or a soil resistant to penetration by water and, usually, by air and roots.

Improvement Cutting

It is process of a cutting made in a stand, which has passed the sapling stage primarily to improve composition and quality by removing less desirable trees of any species.In felling a tree, the initial cut that removes a wedge-shaped piece of wood and determines the direction of fall.

in situ

It refers to the seed planted in the same area as it was collected (On the site).

in vitro

The term refers to a laboratory condition (strictly 'in glass', that is, a test tube). Eg. the digestibility of animal feeds may be estimated by appropriate chemical methods '*in vitro*' in a laboratory.

in vivo

The term refers to the condition 'In the living body'. Eg. *in vivo* (in the animal itself) using a fistula (narrow tube) to extract material ingested after it has been through the rumen (first stomach), or later, to see what has been absorbed; or, ultimately, by liveweight gain per amount of intake.

Inbred Line

In cross pollinated crops a line produced by continued inbreeding is called as inbred line.

Inbreeding

The mating of individuals more closely related than individuals mating at random for enough generations to make it homozygous.

Income Equivalent Ratio

It is the ratio of the area of sole crops which will give the same gross income as one hectare of intercrop at the same management level.

Incompatibility

The term refers to the inability of gametes to unite and form a zygote. It is due to inability of pollen to fertilize through pollen tube growth, being arrested in the style and due to the inability of embryos to develop.

Increment Borer

The tool that is used to extract a core of wood from a tree, for the study of the radial growth of a tree without felling it.

Increment

It refers to the increase in girth, height, volume, weight or value of individual trees or crops over a stated period.

Indeterminate Flowering

A plant in which flowering is more or less continuous until climatic conditions become unfavourable.

Indeterminate Growth

The term refers to a pattern of development in which the apical meristem remains vegetative, and does not produce flowers, so that new leaves and stems continue to be produced while flowers are also forming.

Indigenous

A particular species which is native to a specified area and not introduced. An indigenous tree is one that grows naturally within a specific environment or within certain predetermined boundaries.

Inductive

The adducing of a number of separate facts, particulars, and similar items, especially to prove a general situation. The opposite of deductive.

Indurated Soil

The soil which is cemented into a hard mass that will not soften on wetting is called as indurated soil and the process in which soil becomes hardened and offers resistance to root growth is called induration.

Industrial Forestry

It is a commercial planting of trees in large-scale for timber and other wood products (for example, wood chips).

Infested

Presence of occurrence of great numbers of insects, mites, nematodes and so on, as applied to an area or field is called as infested condition.

Infiltration Capacity

It is defined as the maximum rate at which water can be absorbed by a given soil per unit surface under given conditions.

Infiltration

Infiltration is the flow of a liquid into a substance through pores or other openings; connoting flow into a soil as opposed to percolation, which strictly means flow through a porous substance i.e. the downward entry of water into soil or other material.

Infiltration Rate

Rate at which a particular soil absorbs water.

Infiltrometer

The device which is used for measuring the rate of entry of fluid into a porous body is called filtrometer.

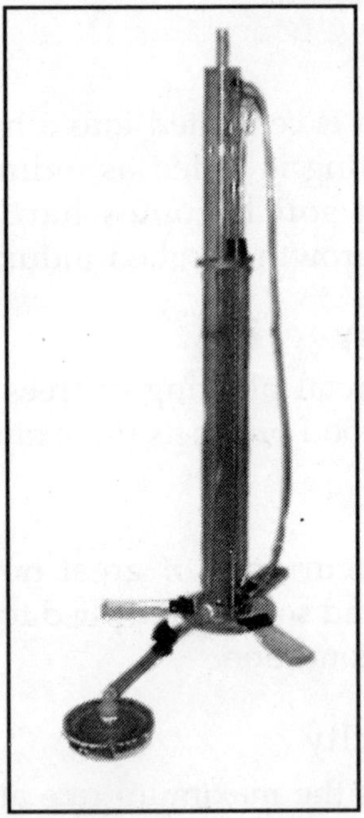

Inflorescence

The system in which flowers are arranged about a plant axis usually set apart from the foliage leaves. The indefinite inflorescence which does not terminate in a flower is called racemose and the one which is definite and terminates in a flower is called cymose.

Informal Survey

Informal survey is the one which includes field study by interviews with farmers, direct observations and existing information which is used to develop an understanding of the farming system's constraints and potential.

Infructescence

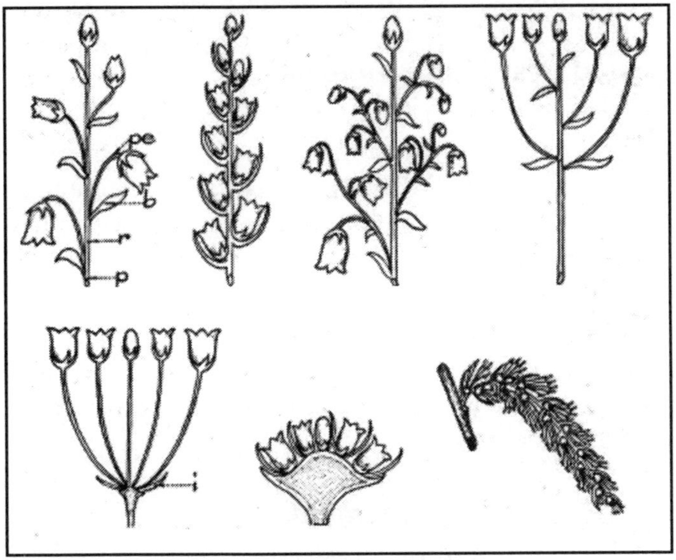

Refers to the inflorescence converted into fruits.

Ingrowth

Ingrowth is defined as the volume or number of trees that have grown past an arbitrary lower limit of measurement during a specified period.

Inoculation

The process of deliberate introduction of material containing microorganisms into soils, nursery composts or living plants is called inoculation.

Inoculum

Inoculum is a preparation containing live micro-organisms (rhizobia, mycorrhizal fungi) that can be used to add to the soil, or directly to the seeds, so as to inoculate suitable plant species with these favourable organisms.

Insecticides

Insecticides are agents or chemicals used to kill insects.

Integrated Pest Management

It is an approach that attempts to use all available methods of pest control ranging from biological means to pesticides with the main objective to keep the damage below economic levels without eliminating the pest completely

Integrated Resource Management

It refers to a form of forest or woodlot management in which a number of resources or roles (e.g. hydrological functions), are obtained or fulfilled at the same time over the whole area.

Intensive Forest Management

The management of forest intensively by basic forest management, juvenile-stand improvement and acceleration of artificial regeneration.

Intensive Silviculture

It refers to the silvicultural practices in established stands (past free-growing condition) to enhance stand value and yield by using genetically improved planting stock; intensive site preparation, such as spraying herbicides to reduce competing vegetation before mechanical preparation and manual weeding of plantations at early stages.

Interaction

Interaction is the tendency for the effect of varying one factor to depend on the level of another factor and it is measured as the difference between two differences.

Intercepted Rainfall

The term synonymous to interception loss refers to the proportion of the gross precipitation that is held up in the canopy and that does not reach the ground either through stemflow or as throughfall and then evaporates back to atmosphere.

Interception

Defined as the process by which precipitation is caught and held by vegetation (or structures), then lost by evaporation without reaching the ground.

Intercropping

Intercropping is the cultivation of two or more crops simultaneously on the same field. If intercropping is done in row arrangement it is called as row intercropping and if

done without a row arrangement it is called as mixed intercropping.

Interculture

Refers to a form of multiple cropping where perennial (usually tree) crops are grown in association with other species to form a multilayered system.

Interference

Interference is a blanket term used to describe the different kinds of competition or sometimes to describe competition only between genotypically identical or similar plants. It also refers to the hardships caused by the proximity of neighbouring plants.

Intermediate Tolerance

It is a characteristic of certain tree species that allows them to survive, though not necessarily thrive, in relatively low light conditions.

Intermediate Trees

Intermediate trees are trees shorter than those in the dominant or codominant classes with crowns either below or extending into the crown cover formed by codominant and dominant trees. They receive a little direct light from above, but none from the sides; usually with small crowns, considerably crowded on the sides.

Intermediate

Trees with crowns extending into the lower portion of the main canopy of even-aged groups of trees, but shorter in height than the codominants. Here, the trees receive little direct light from above and none from the sides.

Intermittent Growth

It is a non-continuous growth without the predictable regularity of rhythmic growth.

Internal Rate of Return

Internal rate of return is the discount rate that, when applied to the streams of costs and revenues generated by a project, will reduce the net present value to zero.

Internode

It is defined as the region of the stem between two succesive nodes.

Interplant

The operation to plant seedlings among existing trees, planted or natural is called interplant.

Interpolated Cropping

It refers to the situation where the crop duration of species is very different, cropping sequence where two or more species are grown on the same unit of land but at least one species has a later sowing (planting) time and an earlier harvest than the other(s).

Intertropical Convergence Zone

It is the dividing line between the northeast and southeast trade winds.

Intimacy

The degree to which various plant component species are spaced so as to be close to one another in intercropping is called intimacy.

Intolerance

It is a characteristic of certain tree species that does not permit them to survive in the shade of other trees.

Introduced Species

Introduced species refers to a nonnative species that was intentionally or unintentionally brought into an area by humans.

Introgression

Introgression is the term which refers to long, continued hydridization leading to the spread of genes from one race or species to another for various traits like resistance to abiotic and biotic stresses.

Iron Pan

It refers to a soil in which iron compounds have been washed from the upper layers and deposited as a hard layer lower down. It prevents free drainage and root penetration but can be broken mechanically by subsoiling.

Irrigated Lowland

It is the availability of irrigation water may be year-round or seasonal.

Isocline

Isocline is a line on a map or chart joining locations having equal slope values.

Isoenzyme Analysis

The analysis of plant proteins by substituting known enzymes in a plant substrate to determine whether a reaction will occur which is done so as to determine the presence of various enzymes within plant tissue.

Isogenic Lines

Isogenic lines are lines which are identical to each other in all loci except at one loci. Whereas clones, homozygous lines, identical twins are identical at all loci.

Isohyet

Isohyet is a line on a chart or diagram drawn through places or points having the same rainfall or precipitation over a stated period.

Isolation

Isolation is the prevention of crossing among populations by various factors like distance or geographic barriers (geographic isolation), growth on different sites (ecological isolation), difference in flowering time (phenological isolation) or genic or chromosomal differences preventing normal seed set (genetic isolation) which is done to preserve the purity of the hybrids.

Isopleth

Isopleth is a line on a chart or diagram drawn through places or points having the same value of a meteorological element.

Isoquant

Isoquant is a line on a graph representing a particular (stated) value of output as the result of interaction of two variables represented on the *x* and *y* axis.

Isotherm

Isotherm is a line on a chart or map passing through points of equal temperature.

Isozyme

The term refers to the different molecular forms of the same enzyme.

It is the outcome arising from those processes brought about by positivie interactions between proximal plants.

Iterative

Iterative is a term which applies to a computational procedure in which replication of a cycle of operations produces results that approximate the desired result more and more closely. But in the context of social science tools, this describes the need to apply tools repeatedly to obtain validity and reliability, and to update solutions.

—xxx—

J

Juvenile Leaf

Juvenile refers to a leaf of a young plant or shoot from the base of an older. Eg. eucalyptus, the first leaves formed after the cotyledons and before the formation of subsquent types of leaves.

Juvenile Wood

Juvenile wood is an inner layer of xylem surrounding the pith, in which the cells are smaller or less structurally developed than those of the outer xylem. The period during which it is formed is termed the juvenile period and this period varies between individual species and environmental conditions.

Juvenility

It is a phase of plant development prior to the time when the plant has the capability to flower.

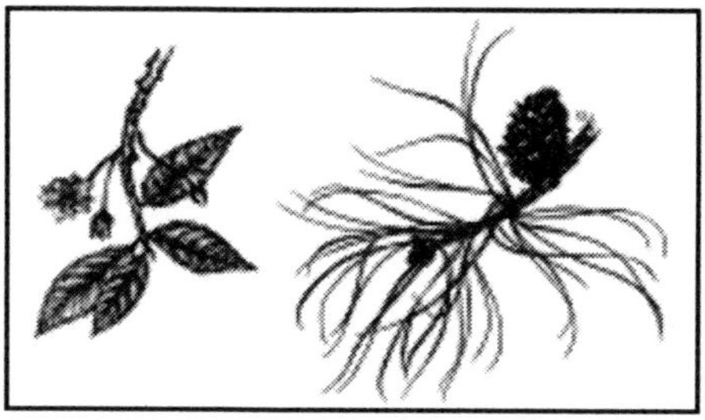

—xxx—

K

Kaolinite

It is an aluminosilicate mineral of the 1.1 crystal lattice group.

Kerf and Saw

The width of a cut made by a saw is called as kerf.

Kiln and Dry

Kiln is a structure heated by gas, steam, or electricity, in which lumber is dried under controlled conditions to a desired moisture content.

Kilning

The process of drying wood under controlled temperature and humidity is called kilning.

Knot

Knot is that part of a branch that has been incorporated into the main stem.

Knotty Core

The central core of wood in a pruned tree that contains knots is called as knotty core.

K-Selection

K selection refers to the selective conditions that favour longevity of individuals at expense of reproductive capacity. The species that is a adapted to maximum survival of individuals at expense of reproductive capacity is called K-strategist.

—xxx—

L

Lammas Shoots

Extra whorls of branches or leader growth extension late in the growing season.

Land Capability

Land capability is defined as the suitability of land for use without damage. It is an expression of the effect of physical land conditions, including climate, on total suitability for use without damage, for crops that require regular tillage, for grazing, for woodland and for wildlife. Land capability involves consideration of the risk of land damage from erosion.

Land Characteristic

It is an attribute of land that can be measured or estimated and that can be employed as a means of describing land qualities or distinguishing between land units of differing suitabilities for use.

Land Classification

It refers to a means of referring to systems of grouping land on the basis of its physical characteristics (including vegetation) without explicit assessment of suitability for kinds of land use.

Land Evaluation

The term refers to the process of assessment of land performance used for specified purposes.

Land Facet

Land facet is an area within which, for most practical purposes, biophysical conditions are uniform. Land facets often occur in regular sequences, within a land system.

Land Race

Land race is a genetically variant population that has originated through selection and propagation by individual farmers, or in small areas of geographic isolation.

Land System

Land system is an area with a recurring pattern of land facets. Initially identified and defined in terms of their land forms, then in terms of the full range of environmental factors involved (eg. 'undulating hills', 'coastal plain').

Land Unit

An area or type of land that posseses relatively homogeneous biophysical characteristics. All land within a land unit has similar resource potential and hazards; it is the basic unit for diagnosis of biophysical resource constraints and potentials.

Land Capability Classification

It is a system for classifying land based on its capacity to support agriculture, grazing, forestry or its usefulness for recreation and conservation and concerned with the fitness of land to support land use, rather than productivity, emphasizing soil erosion.

Land Equivalent Ratio

The land equivalent ratio is the total area of sole crops required to produce the same yields as would be obtained when they are intercropped.

Landing

It refers to a cleared area within a timber harvest where harvested logs are processed, piled, and loaded for transport to a sawmill or other facility or the area where logs are collected for loading.

Leader

The growing top (terminal shoot) of a tree is called as leader. The distance up the main stem of the tree between each whorl of branches generally represents I year of height growth.

Land Suitability Rating

It refers to the partial suitability of a land unit for a land utilization type, based on one land quality, or a partial set of land qualities. Here land suitability ratings are combined to give a land suitability class.

Lapse Rate

Defined as measure of the decrease in temperature per unit increase in vertical height in the atmosphere (approximately 0.6°C per 100 m).

Lateral Bud

It refers to any bud that develops on the side of a stem.

Laterite

Laterite is an iron-rich subsoil layer found in some highly weathered humid tropical soils that, when exposed and allowed to dry, becomes very hard and will not soften when rewetted. Here the erosion removes the overlying layers it exposes the laterite and results in virtual pavement. It is characterized by low base exchange capacity, low clay activity, low content of most minerals and high stability.

Latex

Latex is a white or whitish liquid varying in thickness which is produced by certain plants. For example, *Antiaris toxicaria* and *Bridelia micrantha* both have a latex sap. A more popular example is the rubber tree *(Hevea braziliensis)*. Some types of latex can be harmful, especially if the latex gets into the eyes.

Latin Square

It is an experimental design that attempts to remove two sources of heterogeneity (row and column effects) from residual variation. A 5 x 5 square consists of 25 plots in 5 rows with 5 plots per row, and each treatment occurs once in each row and column.

Lattice Design

It is an incomplete block design (often used for variety or provenance trials) in which the number of treatments must be an exact square.

Layer

A definable horizontal part of a plant canopy. In plant propagation, a shoot that puts out roots at the nodes when it comes into contact with the soil, naturally or artificially.

Layering

The production of adventitious roots from a prostrate main stem or a branch in contact with the ground, enabling a potentially separate plant to be formed.

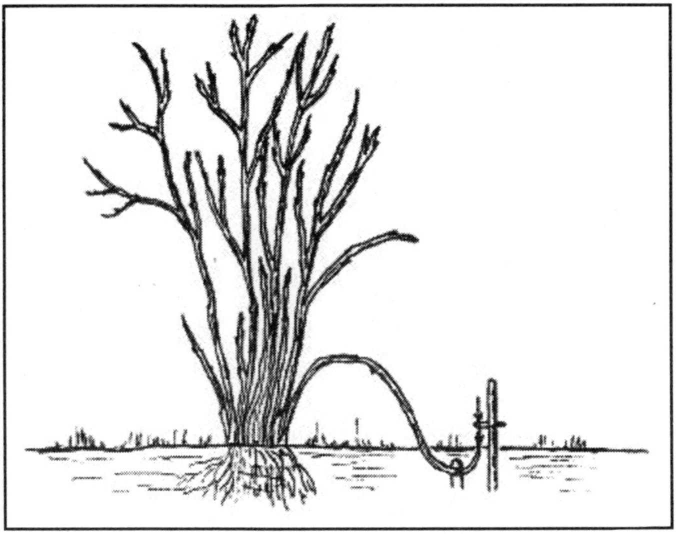

Leached Soil

A soil where most soluble materials have been removed from the entire profile or removed from the layer and accumulated in a lower layer.

Leaching

The washing out of material from the soil, both in solution and in suspension is called leaching.

Leaf Area Duration

It is defined as the Leaf area index integrated over time.

Leaf Area Index

The area of crop leaf per unit area of ground covered by the crop. Can also be estimated for single plants or hedgerows.

Leaf Area Ratio

The ratio of the area of the assimilatory material of a plant to the total weight of the plant. Usually estimated as a mean leaf area ratio over the same period of time being used to examine net assimilation rate. Leaf area ratio = specific leaf area x leaf weight ratio.

Leaf Retaining

The term applies to the plant whose leaves remain for more than one season.

Leaf to Total Growth Ratio

The term refers to the change in leaf area in terms of change in total plant (dry) weight per unit of time which indicates the extent to which a plant is investing its dry matter production in leaf as distinct from non-assimilatory parts.

Leaf Water Potential

The total potential for water in a leaf, consisting of the balance between osmotic potential (arising from the presence of solutes), turgor pressure (the hydrostatic pressure developed in the cell in response to the osmotic potential and the tensile strength of the cell wall) and the matric potential (arising from the imbibitional forces of colloids and of capillaries in the cell wall). This determination of water deficit in leaves can be measured with a 'pressure bomb' or by psychrometric methods.

Leaf Weight Ratio

Leaf weight ratio is the ratio of the weight of leaves compared to the total plant weight, usually averaged over a period.

Leave Trees

It is a natural method of reforestation where trees are left in or just outside a harvest zone (often otherwise a clearcut) to reseed the area. It is often slower, and it does not have the more assured results of direct seeding or planting.

Logger

A person who is engaged in a logging operation or one who moves logs to landings or skidways.

Log Rule

Log rule is a table showing the estimated or calculated amount of lumber (in board feet) that can be sawn from logs of given length and diameter. 1. **Doyle rule -** A simple formula used to underestimates the yield from small logs and over estimate with logs over 28 inches in diameter. 2. **Doyle-Scribner rule.** A combination rule, derived by using Doyle rule values for logs up to 28 inches in diameter and Scribner rule for logs larger than 28 inches. 3. **International rule.** A formula allowing Y2-inch taper for each 4 feet of length and Vs -inch shrinkage for 1-inch board. In one form, it assumes a V8 -inch kerf; in modified form, it assumes a V4 -inch kerf. 4. **Scribner rule.** A diagram rule, one of the oldest in existence. It assumes I -inch boards and V4 -inch kerf, makes a liberal allowance for slabs, and disregards taper. 5. **Scribner decimal C rule.** The Scribner rule modified by rounding off the last digit to the nearest 10 and dropping the zero'. Zeroes are added to total of volumes.

Lee Side

Lee sidw is the downwind, sheltered side.

Legume

Legume is a pod-bearing member of the Leguminosae family, one of the most important and widely distributed plant families. It includes many valuable food and forage species, such as peas, beans, peanuts, clovers, alfalfas, sweet clovers, lespedezas, vetches and kudzu. Some of the plants are nitrogen fixing through nodule formation.

Lenticel

It is an organ that replaces stomata in the structure of the stem which is clearly visible and looks like a little nodule.

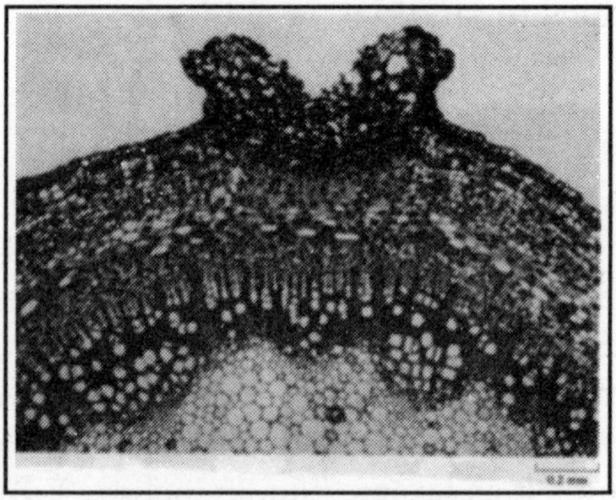

Level

A particular amount or category of a factor, for example, in a fertilizer rate trial the levels of the factor nitrogen may be nil, 50, 100 and 200 kg of N per hectare. In a pruning and thinning trial the factor levels will be intensities and times of pruning and thinning. Different methods of cultivation, types of plant container, brands of herbicide and times of spraying are other examples of 'levels'.

Level Terrace

A terrace that follows the absolute contour, as contrasted with a graded terrace. Used only on permeable soils where conservation of moisture for crop use is particularly important or where outlet channels are impractical.

Ley Farming

The system which follows rotation of arable crops with two or more years of sown pasture.

Ley Pasture

A pasture that is temporary and grown as a specific phase in a defined crop rotation sequence is called as ley pasture.

Liana

Vinelike climbing plant that grows by supporting itself on others and that roots in the ground, germinating there and maintaining its contact with the soil.

Lift

Lift is a stage in the sawlog pruning regime for a tree; usually three separate lifts are needed to ensure a cylindrical knotty core without taper.

Lift Pruning

The process of removing the lower branches of trees being grown for timber, to elevate the crown and to obtain a better quality, knot-free log.

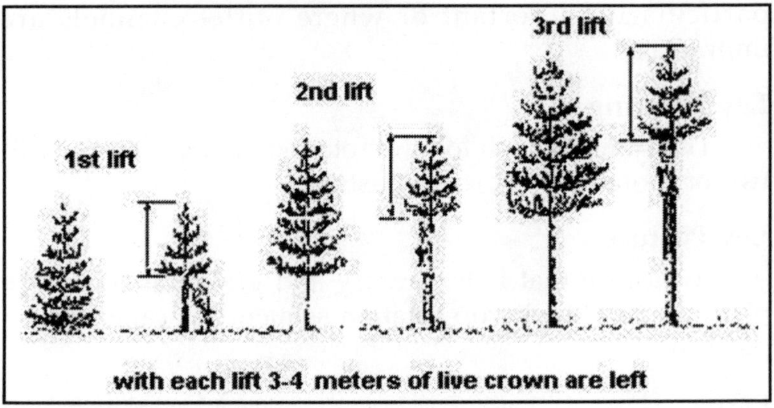

Light-Use Efficiency

It is defined as the rate at which a stand (that is, a unit area of vegetation) produces dry matter (or a harvested part) as a function of solar radiation received. It is also referred as dry matter to solar radiation ratio.

Lignin

Lignin is a 3-dimensional polymer (based on hydroxylated cinnamyl alcohols, for example) that is interspersed in networks of two-dimensional cellulose molecules forming micelles (small aggregations) in plant cell walls, which are thereby greatly strengthened. Lignified tissues decompose relatively slowly and give rise to humus.

Lignotuber

A woody organ arising from an accessory meristem in the axils of cotyledons or leaves near the base of a stem and having a proliferation of dormant buds or a swollen, woody structure at the base of many eucalypts (particularly 'mallees') containing dormant buds.

Lime

In strict chemical terms, calcium oxide. In practical terms, a material containing carbonates, oxides or hydroxides or both that are used to neutralize soil acidity is called lime.

Line Breeding

A breeding system in which a number of genotypes that have been progeny tested in respect to some character or group of characters are combined to form a variety is called as line breeding. It can also defined as a system of mating in which closely related individuals are crossed with each other.

Line

Line refers to a population consisting of one (self line), a few (inbred line), or many (bulked line) parents and their offspring through several generations.

Line Planting

Line planting is a process of setting trees in parallel rows, generally at regular intervals between and in lines, on land either wholly or partially cleared. A form of line planting sometimes known as **corridor planting**, involves setting a line of trees in narrow lanes ("corridors") that cut through undergrowth at more or less regular intervals (sometimes at their final crop spacing); generally a form of improvement planting or enrichment.

Line Thinning

The process of removing specified rows of trees in a plantation, such as every fourth row is called as line thinning.

Linear Data

Linear data is defined as a data swarm whose projection onto any two-dimensional space, however oriented, gives a two-dimensional swarm with a long axis that is (approximately) a straight line. The data are nonlinear If any projection yields a swarm with a curved axis.

Linear Transformation

The term refers to changing the basis of the values of a data set describing the association of two variables so that when plotted they form a straight line.

Litter

Litter is the uppermost layer, on the soil surface, of loose organic debris that consists of freshly fallen or slightly decomposed organic materials.

Live Fence

The term living fence refers to the way of planting a line of trees and/or shrubs (the latter usually from large stem cuttings or **stumps**), at relatively close spacing and by fixing wires to them for establishing a boundary. More uprights (dead sticks) can be tied to the wires if animals are to be kept out.

Live-Crown Ratio

A rough but convenient index of the ability of a tree's crown to nourish the remaining part of the tree; it is the percentage of length of stem having living branches.

Livestock System

A system in which a land unit comprising pastures, herds and auxiliary feed sources, transforms plant biomass into animal products.

Livestock Unit

The term refers to a unit used in animal husbandry based on the amount of intake required for a mature Friesian cow (taken to be an average of 550 kg liveweight) and based, experimentally, on $Y = 0.0234x \times 1032$ (where Y = the dry matter intake in kg day–1 and x = the liveweight of the animal in kilograms). Smaller animals (for example, sheep and goats) are assessed as appropriate fractions of a livestock unit. Sometimes called as 'cow equivalent'.

Loam

The soil containing a relatively equal mixture of sand and silt and a somewhat smaller proportion of clay; generally a desirable quality. It is subdivided into texture classes like 'sandy loam', 'silt loam' and 'clay loam'. The composition of loam is 7 to 27% clay, 28 to 50% silt, and less than 52% sand.

Loblolly-Shortleaf Forest Type

An association of tree species common to the southeastern United States that includes loblolly and shortleaf pines and oaks.

Lodging

Lodging is the collapse of top-heavy plants, particularly grain crops, due to factors like excessive growth or beating by rain.

Loess

Mainly silt, transported and deposited by wind is called as loess.

Log

It is the part of the trunk which is suitable for use as timber.

Log Rule

It is a method for calculating wood volume in a tree or log by using its diameter and length. The international 1/4-inch rule is the legal rule in Maryland.

Log Scale

It refers to the lumber content of a log as determined by a log rule.

Lookout Fire spotter

A station or post used primarily in the detection of fires, often an observation tower located on a high point of ground.

Lop

The term applies to chop branches, tops, or small trees after felling, so that the slash lies close to the ground or to cut the limbs from a felled tree.

Long Shoot

Where there is shoot dimorphism, an extended shoot contributing to the architecture of a tree is called as long shoot.

Long-Day Plant

Long day plants are plants whose development is affected by photoperiodism, in particular where a process (for example, flowering) is promoted if the plant is subjected to day lengths above a critical length and night lengths below it.

Longevity

The term refers to long life or long duration of existence.

Lopping and Scattering

The process of lopping the slash created after felling and spreading it more or less evenly over the ground without burning.

Lopping

The technique of cutting one or more branches of a standing tree to collect fuel or fodder for animals by not disturbing main stem is called as lopping.

Lot

A plot of land is called as lot.

Low Thinning

The removal of trees from the lower crown classes in a stand.

Lumber

Sawn timber. A unit of measure is the 'board foot', that is, a board 12 inches long by 12 inches wide by 1 inch thick.

Lump-Sum Sale

The term refers to a timber sale in which an agreed-on price for marked standing trees is set before the wood is removed (as opposed to a unit sale).

Lysimeter

A vessel or container placed below the ground surface to intercept and collect water moving downward through the soil in which vegetation can be planted, and which is isolated hydrologically from the soil around it. Lysimeters are used to study various aspects of the hydrological cycle.

—×××—

M

Macroclimate

The climate of a region.

Macronutrient

A chemical element necessary in large amounts (usually > 50 ppm in the plant) for the growth of plants. Includes C, H, O, N, P, K, Ca, Mg, S. 'Macro' refers to quantity and not to essentiality of the element.

Main Effect

The response to different levels of a treatment factor, measured independently of other factors. In a balanced design such as a randomized complete block, the main effects of levels of a treatment factor are estimated by averaging the corresponding responses over all levels of every other factor.

Mainline

The term refers to different meanings i.e. a). To chop branches, tops, or small trees after felling, so that the slash lies close to the ground, b). To cut the limbs from a felled tree, c). In cable logging, the line used to retrieve turns of logs, d). The main access road to a forest tract.

Management Boundary

It is a basic spatial unit suggested to embrace agriculture, horticulture, forestry and agroforestry; an ecologically homogeneous area set aside for a particular use.

Management Plan

Management plan is a written plan for the organized handling and operation of a forest property. It usually includes data and prescribes measures designed to provide optimum use of forest resources according to the landowner's objectives.

Mangroves

Mangroves refers to the open or closed stands of trees and bushes occurring in the tropics in intertidal zones, usually around the mouths of rivers, creeks and lagoons where soils are heavy textured and have a fluctuating salt content and soil level. A name used collectively for the assemblage of plants, as well as to refer to individual genera or species, for example, woody members of the Rhizophoraceae, Combretaceae and Verbenaceae. These have knee-like roots that are 'pneumatophores' (roots with air spaces). Grown for wood and tannins.

Marking Timber

Selecting and indicating, usually by an axe mark (blaze) or paint mark, trees to be cut or retained in a harvesting operation.

Manure

The animal excreta with or without a mixture of bedding or litter is called as manure.

Marginal Land

Land that does not consistently produce a profitable crop because of infertility, drought, or other physical limitations such as shallow soils.

Marker Gene

A gene, usually dominant, with a large effect and helps to identify one parent in offspring resulting from open pollination.

Marketing

Marketing is the selling of timber or other forest resources. Satisfactory price is got through competition, skillful negotiation, knowledge of timber markets, and the aid of a competent broker or consultant by the sellers.

Mass Selection

A form of selection in which individual plants are chosen and the next generation propagated from the aggregate of their seeds (with self-pollinated plants) based on their phenotype.

Mast

Mast refers to the fruits or nuts used as a food source by wildlife. Soft mast include most fruits with fleshy coverings, such as persimmon, dogwood seed, or black gum seed. Hard mast refers to nuts such as acorns and beech, pecan, and hickory nuts.

Mast Fruiting

It refers to a period of massive fruit production.

Matrix

Matrix mathematically refers to a two-dimensional array of numbers. The meaning of each number (or element) depends on its position in the matrix, that is, on the row and the column in which it occurs. It also refers to the generally rigid and connected material in which other matter is retained.

Mature Tree

A tree that has reached a desired size or age for its intended use. Size, age, or economic maturity varies depending on the species and intended use.

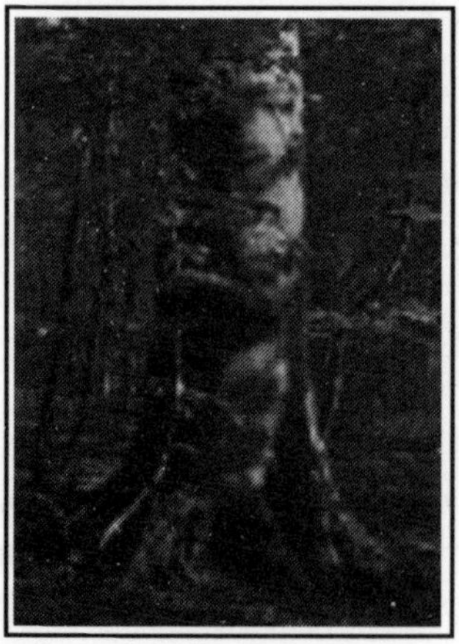

Maturity

The phase in the development of a plant which is usually irreversible that marks the transition from **juvenility,** when the plant is unable to flower, to a full capacity to form flowers that are sexually reproductive unless the plant shows apomixis.

MBF

Abbreviation denoting 1,000 board feet. MBF is a typical unit of trade for dimension lumber and sawtimber stumpage. (It takes 11 MBF of wood to build an average 1,900-square-foot house.)

Mcleod

Firefighters also use this combination hoe, rake and scraping tool and rake to remove plants and shrubs when building a fireline.

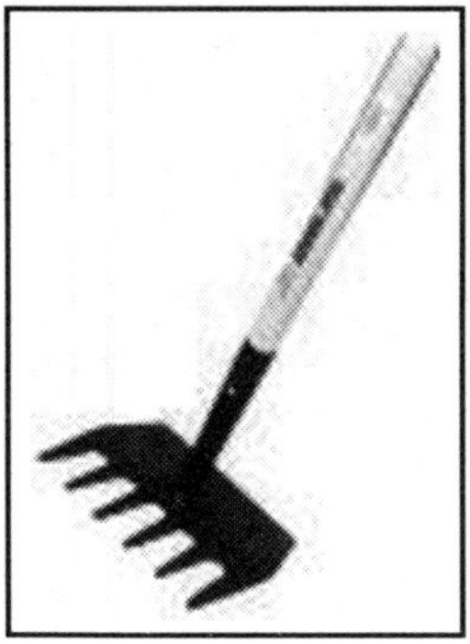

Mean

A measure of central tendency, for example, a 'sample mean'. The 'arithmetic mean' is obtained by dividing the sum of the observations by their number; the 'geometric mean' is the arithmetic mean of the logarithms.

Mean Annual Increment

In forestry, for a particular stand, the total increment of wood up to a given stand age (in years) divided by that age. The mean annual increment for a whole rotation is termed the final mean annual increment.

Mechanical Thinning

It is a process of thinning in which a predetermined proportion of the trees is removed without regard to size or quality, such as, for example, every third tree.

Mechanistic Model

A model based on known mechanisms that operate within the model, such as the fundamental laws of physics and chemistry. It may be a subclass of a deterministic model.

Median

A measure of central tendency. The middle value in a frequency distribution, that is, as many values lie above as below it.

Mensuration and Forest

Mensuration is defined as the science dealing with the measurement of the volume, growth, and development of individual trees and stands, and the determination of the various products obtainable from them.

Merchantable

The term refers to that part of a tree that can be manufactured into a salable product.

Merchantable Height

The length of the tree stem from the top of the stump to the top of the last merchantable section. Usually expressed in ft or number of logs.

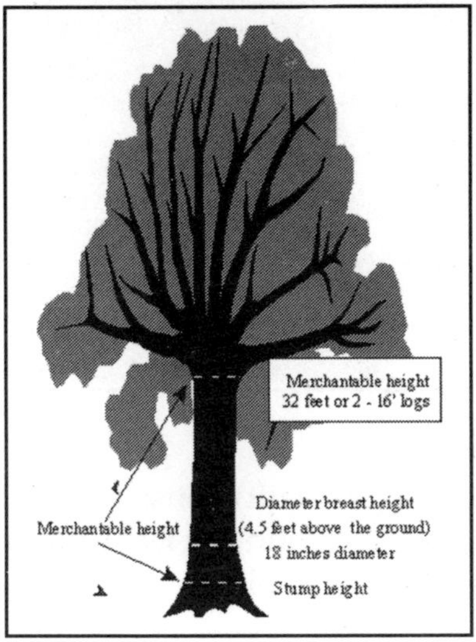

Merchantable Timber

A tree or stand of trees that may be converted into salable products.

Merchantable Volume

The amount of wood in a single tree or forest stand that is considered salable is called as merchantable volume.

Meristem

A plant tissue composed of dividing cells and giving rise to organs such as leaves, flowers, xylem, phloem, roots. A region of a plant in which cells are not fully differentiated and are capable of repeated mitotic divisions.

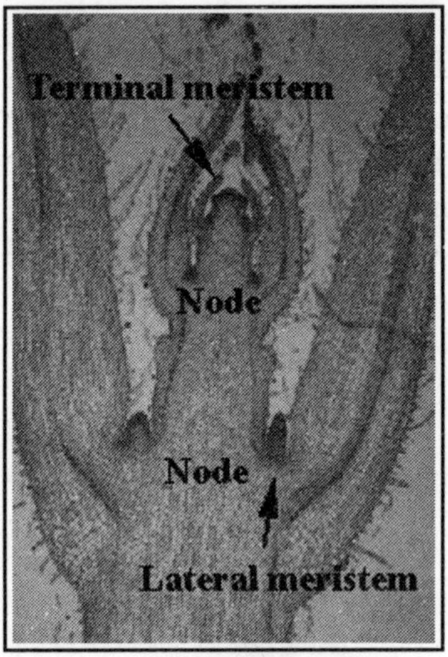

Meristem Culture

A part taken (an explant) consisting only of the apical dome of tissue above the youngest leaf primordium for tissue culture to regenerate new plants.

Mesophyte

A plant that requires a moderate amount of water for normal growth is called mesophyte.

Metabolism

The process or sum total of chemical reactions by which cells or organisms utilize nutritive material to build living matter and structural components, or to break down cellular material into simple substances to perform special functions.

Metabolizable Energy

The term applies to the usable energy in animal feeds taking into account the amounts of different constituents (carbohydrates, proteins etc,) their digestibility and their calorific value.

Micro Water Catchment

It refers to a part of water harvesting i.e. a modification to the soil surface on a slope which traps and conveys runoff water to a small pit, adjacent to which a tree is planted.

Microclimate

Microclimate is the climate as found in small areas, for example, around plants.

Microfauna

The small animals that can be seen only with a microscope, such as protozoa, nematodes.

Microflora

The small plants that can be seen only with a microscope, such as algae, fungi.

Micronutrient

The chemical element that is necessary in only extremely small amounts (< 50 ppm in the plant) for the growth of plants. Examples are B, Cl, Cu, Fe, N, Zn. 'Micro' refers to the amount used rather than to essentiality. Also called 'trace elements'.

Microorganism

A member of the microflora or microfauna that can be seen only with a microscope.

Micropropagation

Various procedures by which small pieces of plant tissue are taken, surface sterilized and then grown on, or in, an appropriate medium under carefully controlled environmental conditions and provided with appropriate nutrients, growth hormones, and supplementary components, to allow cells to differentiate into small plantlets (explants)and eventually hardened off before being planted out.

Microsite

The ultimate unit of the habitat, i.e., the specific spot occupied by an individual organism. By extension, the more or less specialised relationships existing between an organism and its environment.

Mimosine

An alkaloid toxic to animals, which is contained in some members of the legume family (Mimosaceae) including *Leucaena leucocephala.* Its toxicity minimizes the amounts of leafy material that can be fed, although **ruminants** can detoxify mimosine through bacterial action in the rumen provided they have the right microflora (either naturally or by inoculation).

Mineral Commodities

A mineral resource with sufficient value to be individually produced and traded on the open market.

Mineral Resource

Mineral resource is a known or undiscovered concentration of naturally occurring solid, liquid, or gaseous material in or on the Earth's crust in such form and amount that economic extraction of a commodity from the concentration is currently or potentially feasible. *Identified, Demonstrated:* An in-place deposit for which there are physical measurements of size and grade combined with limited extrapolation of geological, geochemical, and geophysical data.

Mineral Soil

A soil containing lessthan 20% organic matter or having a surface organic layer less than 30 cm deep is called mineral soil.

Mineralization

The conversion of an element from an organic form to an inorganic state as a result of microbial decomposition is called as mineralization.

Minimum Tillage

Minimum tillage is the crop production where soil cultivation is kept to the minimum necessary for crop establishment and growth.

Miombo

Miombo is a form of woodland found in southern Africa in hot, semi-arid regions with a monomodal rainfall (for example, in Tanzania, Zimbabwe, Malawi, Zambia). Rainfall is high (> 1000 mm) and soils are poor in nutrients and acid. The vegetation is dominated by slow-growing, mainly deciduous, caesalpiniaceous trees like *Brachystegia, Julbernardia*.

Mistletoe Infection Rating

The relative abundance of mistletoe in the crown of a tree or shrub. Is called as mistletoe infection rating.

Mixed Cropping

Growing more than one species on the same piece of land at the same time, or with only a short interval between is called as mixed cropping.

Mixed Farming

The type of farming that involves crop and animal production for betterment of farmers.

Mixed Garden

A land-use form on private lands outside the village, dominated by planted perennial crops, mostly trees, under which annual (seasonal) crops are cultivated.

Mixed Stand

A timber stand in which less than 80 percent of the trees in the main canopy are of a single species.

Mixing Height

The height above the surface (m) defining a boundary layer within which pollutants are free to mix.

Mixing Ratio

It is defined as the ratio of the mass of water vapour to the mass of dry air.

Mode

A measure of central tendency, the value around which there is the greatest concentration in a frequency distribution. It also referred as the most common value (that is, the value with the highest frequency) in a frequency distribution.

Model

A quantitative representation, which, if complex, may require algebraic and arithmetic manipulation. Such models are essential elements of systems analysis, statistical analysis and many forms of computer simulation or model refers to a simplified representation of a system expressed in words, diagrammatically or by mathematical symbols or a simplification of the real world that reveals the key processes necessary for prediction. Models may be verbal, graphic, mechanical or mathematical, depending on their purpose and the need for them; some can be used for quantification. Eg. analytical model, continuous model, deterministic model, dynamic model, empirical model, mechanistic model, numerical model, simulation model, stochastic model.

Monoculture

Monoculture refers to the practice of growing a crop consisting of a single species on a given land area.

Monoecious

Plants bearing unisexual flowers, both staminate and pistillate, on the same plant is called monoecious. Eg. Japanese black pine.

Montmorillonite

A hydrous aluminosilicate clay with 2 : 1 expanding crystal lattice. Expansion on wetting is caused by water moving between silica layers of contiguous units.

Mor

An accumulation of acid organic matter at the soil surface beneath a forest.

Mor Humus

A type of forest humus layer of unincorporated organic material, usually matted or compacted, or both. Distinctly delimited from the mineral soil, unless the latter has been blackened by washing in organic matter.

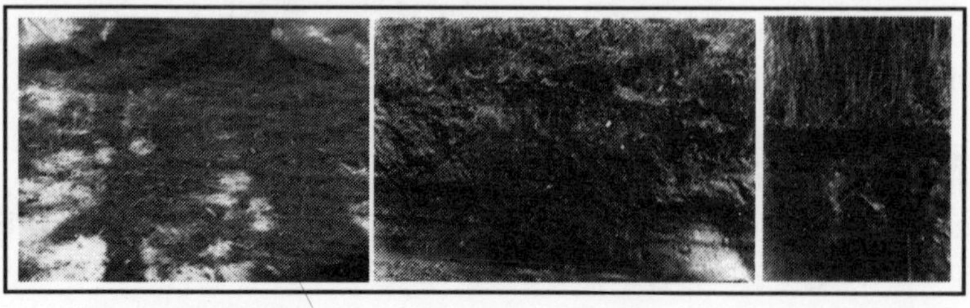

Moraine

Any type of constructional topographic form consisting of till (stiff clay with boulders and sand, and the like) and resulting from glacial deposition.

Morphology

Morphology is a branch of science that deals with form and structural shape.

Mortality

The number or sound wood volume of healthy trees that have died from natural causes during a specified period is called as mortality.

Mosses and Lichens

Mosses are plants growing in tufts or clusters on the ground, decaying wood or on rocks. Lichens are any of the various flowerless plants composed of fungi and algae, commonly growing in flat patches on rocks, trees, etc.

Most Hazardous Pest

The principal natural agent that operates in the vicinity of the sample point and presents the greatest threat to realizing stand goals of stocking, growth and structure is termed as most hazardous pest.

Mountain Beaver

A small nocturnal rodent, found throughout the Coast Range in Oregon and Washington. This burrowing animal has a voracious appetite for Douglas-fir seedlings.

Mulch

Mulch is a natural or artificially applied layer of plant residues or other materials such as stones, sand, paper or **brush** on the surface of the soil or a covering of plant material put on the soil to improve its fertility, moisture retention capacity and organic content.

Mulch Tillage

Tillage or preparation of soil in such a way that plant residues are left on the surface is called mulch tillage.

Mull

A crumbly, intimate mixture of organic and mineral material formed mainly by worms, particularly earthworms is called mull.

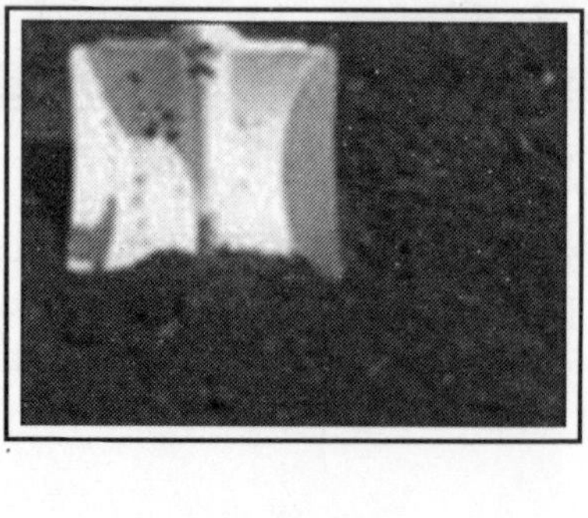

Mull Humus

A type of forest humus layer consisting of organic and mineral matter so mixed that the transition to the underlying mineral layer is not sharp. Associated with conditions of rapid mineralization and organo-mineral complexes of the clay fraction. Mixed microflora, bacteria and actinomycetes predominate.

Multilocational Experiment

An experiment (or trial) conducted at more than one place with a view to including a larger range of **edaphic** (soil) and microclimatic conditions is called multilocational experiment.

Multiple Cropping

Growing more than one crop on the same piece of land.

Multiple Regression

It refers to the relationship between specified values of two or more **variables** (independent variables $x1$ and $x2$) and the expected value of a random variable (dependent variable b) of the form $E(y) = b0 + b1x1 + b2x2$.

Multiple Use

The management of land or forest for more than one purpose, such as wood production, water quality, wildlife, recreation, aesthetics, or clean air is called multiple use.

Multiple-Use Forestry

The concept of forest management that combines two or more objectives, such as production of wood orwood-derivative products, forage and browse for domestic livestock, proper environmental conditions for wildlife, landscape effects, protection against floods and erosion, recreation, and protection of water supplies is termed multiple use forestry.

Multiple-Use Management

Management of land resources with the objective of achieving optimum yields of products and services from a given area without impairing the productive capacity of the site.

Multipurpose Tree

A woody perennial that is purposefully grown to provide more than one significant contribution to the production or service functions (for example, shelter, shade, land sustainability) of the land-use system that it occupies.

Multistage Random Sample

The sample selective scheme with more than one level of random selection; for example, the random selection of villages within a district, then of farmers within each selected village.

Multistorey Cropping

Multispecies crop combinations involving both annuals and perennials with an existing stand of perennials. An association of tall perennials with shorter statured crop species.

Multistorey System

An agroforestry system, such as a homegarden, that has a number of plant components of differing stature so that several layers of canopy are formed.

Multivariate Analysis

A loose term denoting the analysis of data that are multivariate in the sense that the value of each of *k* variables has been measured on each individual. Examples are canonical coordinate analysis and principal components analysis.

Multivariate Factor Analysis

A mathematical method of analysing sets of data to discover which variables are statistically linked, by causative or other factors.

Mutant

Mutant is an individual possessing a new, heritable trait as a result of a process of mutation.

Mutation

Mutation is a sudden, heritable variation in a gene or in a chromosome structure.

Mutualism

An interaction or a relationship between two living beings where interaction is obligatory and favourable to both of them. Eg. mycorrhizae and *Rhizobium* bacteria.

Mycorrhiza

An association that is probably symbiotic (or at least not parasitic) between the root or rhizome of a green plant and a fungus. There are two types i.e. ectotrophic mycorrhiza, endotrophic mycorrhiza.

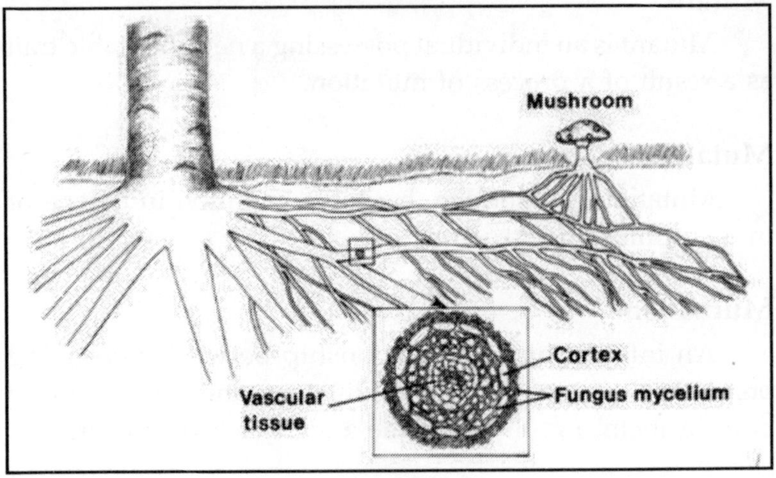

—xxx—

N

National Forest

National forest is the federally owned land managed to provide wood, water, and other uses for the benefit of the people.

National Park

Federally owned land managed to maintain areas of outstanding and unique scenery and geographic features for public enjoyment.

Natural Regeneration

In forestry the term applies to the renewal of forest by self-sown seed or vegetative means. This term may also be used to describe a young tree crop itself.

Natural Resources Conservation Service (NRCS)

The branch of the USDA that coordinates and implements soil conservation practices on private lands.

Natural Selection

It is the elimination of living organisms on the basis of their inability to survive and reproduce under a particular set of environmental conditions.

Natural Thinning

Natural thinning is the death of trees in a stand as a result of competition.

Needle cast

The premature browning and dropping of needles due to fungus infection in Douglas-fir Christmas trees.

Necrosis

Necrosis is defined as the death of (plant) tissue.

Needleleaf Tree

A tree with pointed needle-like or scale-like leaves. Eg. conifers.

Nematocidal Plant

It is a trap crop or a plant species that produces organic substances from the roots which adversely affect (kill) eelworms (nematodes) and sometimes make the larvae unable to infect the plant as it is resistant to them. Eg.*Tagetes* spp.

Neotropics

Neotropics includes both tropical Central and South America.

Net Assimilation Rate

It is the increase in plant material (dry weight) per unit of assimilatory material. Usually estimated as mean net assimilation rate per unit of time (for example, over a week).

It is based on an assumption that there is a linear relationship between increase in leaf area and increase in dry weight.

Net Present Value

The value of a project, used for comparative purposes, derived by discounting both the forecast costs stream and the revenues stream back to the present, at a given discount rate.

Net Radiation

It is defined as the difference between downward and upward (total and terrestrial) radiation i.e. the net flux of all radiations.

Neutral Soil

The soil that is neither acid nor alkaline and having Ph of 6.6 7.3 is called as neutral soil.

Niche

Niche refers to the total range of conditions under which the individual (or population) lives and replaces itself, or the position or status of an organism within its community and ecosystem resulting from the organism's structural adaptations.

Nicking

The practice of cutting into the stem below a bud so as to prevent it from growing out is called nicking.

Nitrification

The process of the oxidation of ammonia to nitrite, and of nitrite to nitrate, by microorganisms is called as nitrification.

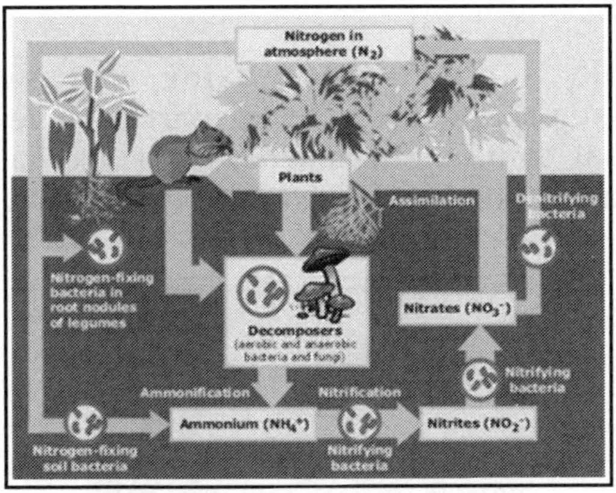

Nitrogen Fixation

The conversion of elemental N to organic combinations or to forms readily usable in biological processes.

Nitrogen-Fixing Plant

A plant that can assimilate and fix the free nitrogen of the atmosphere with the aid of bacteria living in root nodules.

Node

The region of the stem where one or more leaves are attached or the nodes are the horizontal lines linking classes of equal rank. The internodes are the vertical lines linking each class above and below it in rank.

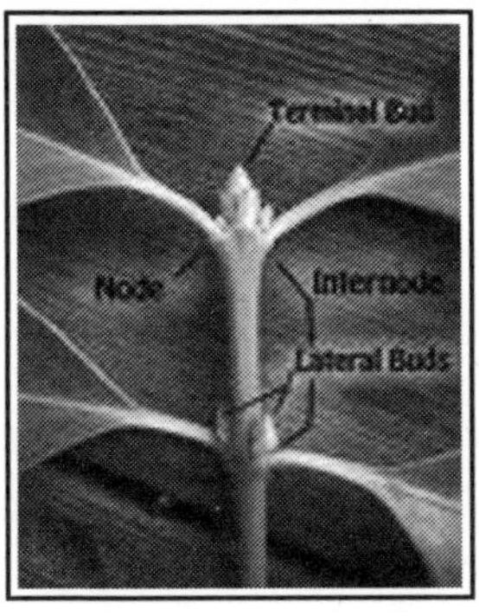

Nodule

Nodule is the nitrogen-fixing root swelling of characteristic shape and size for particular leguminous species that contain rhizobia. If the rhizobial strain is effective, atmospheric nitrogen can be fixed and is readily utilizable by the plant.

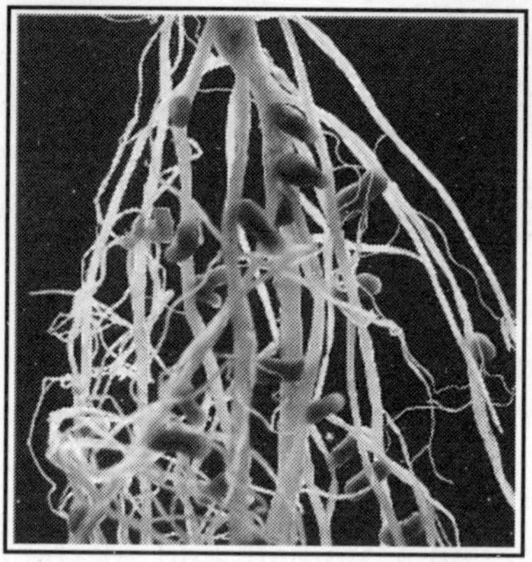

Nomadism

The practice of living by moving from place to place.

Non-Additivity

It is the phenomenon of leading to an expressed interaction.

Nongame Wildlife

Wildlife species that are protected by state wildlife laws and can not be hunted. Eg. Songbirds, eagles, etc.

Nonindustrial Private Forestland (NIPF)

The forestland that is owned by a private individual, group, or corporation not involved in wood processing.

Non-Parametric Method

It includes methods of statistical analysis that does not rely on assumptions about the form of distribution of the analysed variables. Usually they are based on the ranks of the observations rather than their actual values. It is also known as 'distribution-free method'.

Non-Porous Wood

It refers to wood devoid of pores or vessels which is featured by all conifers and a few hardwood species.

Nontidal Wetlands

It refers to the wetlands not affected by ocean tides which are subjected to special regulations.

Normal-Aged Forest

A forest composed of a series of age classes in proportions that permit sustained yield by felling under an appropriate system.

Normalized Data

It is the coordinates of a data point, or the elements of a vector, rescaled so that their squares sum to unity.

Northern Hardwood Forest Type

It is an association of tree species common to the Northeastern United States that includes trees like sugar maple, red maple, yellow birch, hemlock, and American beech.

Notching

Notching is a process of cutting a small wedge from the stem above a bud so as to promote its growth by preventing the flow of inhibitory hormones (indole acetic acid) from the stem apex.

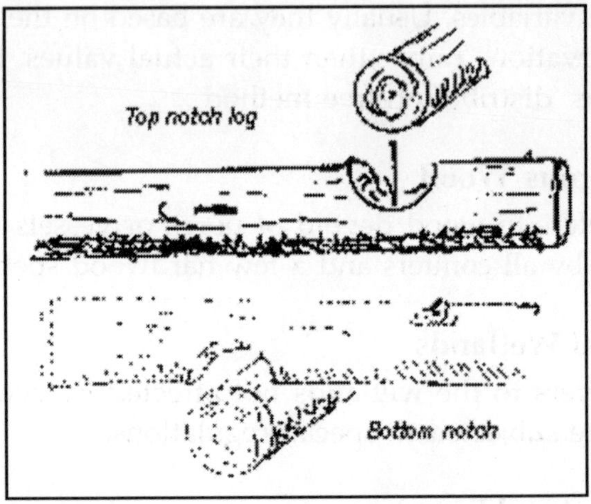

No-Tillage

It is the method of planting crops involving no seedbed preparation other than opening small slits in the soil so that seed can be placed at the intended depth. Generally there is no cultivation during crop production, but chemicals are often used for weed control. Is also called referred as **zero tillage**.

Numerical Model

A model in which the governing equations are solved by means of step-by-step numerical calculations, generally necessitating the use of a computer.

Nurse Tree (Nurse Crop)

A tree or crop of trees, shrubs, or other plants, either naturally occurring or introduced, used to nurture or improve the form of a more important tree or crop during youth by protecting it from factors like frost, insolation, or wind.

Nursery Stock

It refers to shrub or tree species grown in a plant nursery for planting out elsewhere.

Nutritive value

A term usually prefixed by "high," "low," etc., to indicate relative quality of a given forage or feed to furnish elements valuable for animal nutrition.

—xxx—

O

Oak-Gum-Cypress Forest Type

It is an association of tree species common to the bottom lands of the Southeastern United States.

Oak-Hickory Forest Type

An association of tree species common to the Northeastern United States that includes trees like oak, hickory, yellow poplar, and red maple.

Oak-Pine Forest Type

An association of tree species common to the Southeastern United States that includes trees like loblolly pine, Virginia pine, northern red oak, and white ash.

Occlusion

The process of healing of cut branch stubs by the cambium of the surrounding stem surface is referred as occlusion.

Old Growth

A forest that has never been changed by management or harvesting. This term is misapplied by many to describe any forest that appears to be old. Individual trees in this type of forest are usually over 200 years old, and there are large standing and fallen dead trees throughout the stand.

Operation

The term is used for logging jobs, harvesting, cutting, milling, etc. An all-inclusive term for harvesting and hauling out the forest products.

Outplant

Planting nursery-grown tree seedlings on a freshly prepared area is called as outplant.

Overgrazing

Grazing so heavy that it impairs future forage production and causes range deterioration through damage to plants, soil, or both.

Overmaturity

That period in the life cycle of trees and stands when growth or value is declining.

Overrun

Overrun is the excess lumber sawn from logs over the estimated volume or log scale, usually expressed in percent of log scale.

On-Farm Experimentation

It is a generic term to indicate all kinds of scientific experiments that are carried out to evaluate new agricultural technology within the context of existing cropping and livestock systems. Main types are 'on-farm experiments' and 'on-farm trials'.

Open Forest

A '**forest**' where the canopy is not closed and a large proportion of the area does not carry trees.

Open Grown

Trees with crowns which receives full light from all sides due to the openness of the canopy.

Open Loop

Open loop is a control system in which corrective action is not automatic but depends on external intervention.

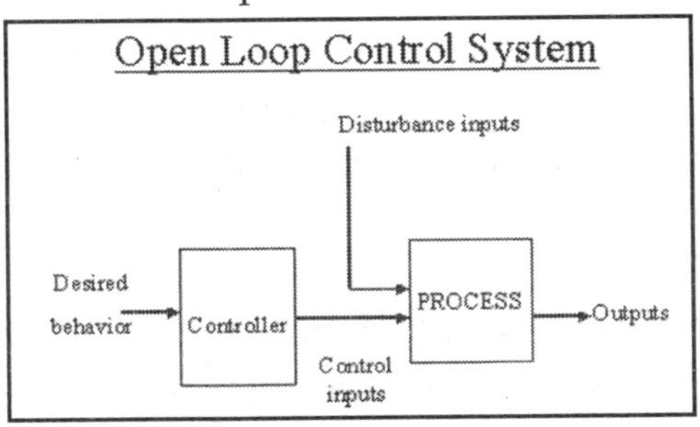

Open Pollination

It refers to uncontrolled pollination brought about by external agents like wind or insects.

Open System

A system which exchanges matter and energy with the surroundings.

Opening Up

Openingup is the considerable reduction of canopy density, e.g., by lopping, felling, or poisoning selected trees, or naturally through pests, disease, or drought mortality.

Opportunity Cost

The term refers to the benefit forgone to an individual or society by using a scarce resource for one purpose instead of for its next best alternative use.

Ordination

A method for reducing the dimensionality of sets of data with as little loss of information as possible, so that the configuration of the data sets can be examined in various ways. For example, principal components analysis can be used statistically or for pattern analysis.

Organic Farming

The production of crops from land that does not receive and has not received (for a stated period of time) any inorganic inputs (fertilizers, pesticides, and so on).

Organic Matter

A general term for plant and animal material in or on the soil and in all stages of decomposition.

Organic Soil

Soil containing a high percentage of organic matter (> 20%) is called as organic soil.

Origin

For an indigenous stand of trees, the place in which the trees are growing. For a non-indigenous stand, the place from which the seed or plants were originally introduced.

Ornamental

Ornamental refers to any tree or plant that is planted for its beauty. This does not mean that it has no other uses. Eg. *Chlorophora excelsa* planted as an avenue tree along roadsides has timber value.

Ortet

Ortet is the original plant from which a clone is started through rooted cuttings, grafting, or tissue culture, or other means of vegetative propagation. The original plus tree used to start a grafted clone for inclusion in a seed orchard is the ortet.

Orthodox Seed

The seed that can be dried and stored at subzero temperatures without damage and with beneficial effects on longevity is called as orthodox seed.

Orthogonal

Orthogonal refers to factors whose effects are estimated independently in an experiment.

Orthotropic Branching

The term applies to type of branching where stems that only grow vertically.

Orthotropy

The gravitational response which produces a vertical axis that result in radial symmetry and vertical orientation.

Outbreeding

Outbreeding is a system of producing progeny sexually that involves the frequent exchange of genetic material between individuals of a population. Outbreeding plants may be self-infertile or have mechanisms to ensure outcrossing.

Outcross

A cross, usually natural, to a plant of a different genotype.

Oven-Dry Soil

Soil that has been dried at 105°C until it reaches constant weight is called as oven dry soil.

Over Bark

Measurement of the diameter or circumference of a trunk or branch made with the bark left on.

Overall Application

Where a fertilizer or herbicide is applied uniformly over the whole crop area, as opposed to band application.

Overhead Application

The application of a herbicide over the whole crop area, as differentiated from application directed specifically to weeds is called as overhead application.

Overlapping Cropping

A cropping sequence where two or more species are grown together on the same unit of land and where the sowing (planting) times do not coincide and the harvesting of one extends beyond that of the other(s); Eg. Relay cropping.

Overstocked

A condition of the stand or forest, indicating more trees than desired, normal, or full stocking would require.

Overstorey

The portion of trees in a forest stand forming the upper crown cover is referred as overstorey.

Overstory Removal

A final harvest in which the cutting releases advance regeneration is called as overstory removal.

Overstory

The portion of the trees is a stand forming the upper crown cover is called overstory.

Overtopped

The situation in which a tree cannot sufficiently extend its crown into the overstory and receive direct sunlight.

Overtopped trees that lack shade tolerance lose vigor and die.

Overtopped Tree

Trees with crowns entirely below the general level of the overstory cover, receiving no direct light either from above or from the sides is called overtopped tree.

—xxx—

P

Paddock

A grazing area that is a subdivision of a grazing management unit, and is enclosed and separated from other areas by a fence or barrier

Palatable

Having a good or at least a tolerable flavour. Some plants, although not poisonous, simply do not taste good and animals avoid them, for example, the leaves of *Prosopis juliflora*. Others are so highly palatable that establishing the tree can become difficult, for example, *Leucaena leucocephala*.

Palm

A typical agroforestry tree belonging to the member of the Aracaceae (previously Palmae) family, containing over

200 genera and more than 2700 species, many of them economically important for food, fibre, canes, waxes, wood, thatch, and so forth.

Pan Coefficient

Ratio of evaporation from a large body of water to that measured in an evaporation pan is called as pan evaporation.

Panmictic Unit

A local population in which there is completely random mating.

Panmixia

Random mating without restriction.

Pantropical

It refers to the tropics as a whole. That is between latitudes 23°27' north and south.

Parallel Row Design

A type of systematic spacing design in which the area per plant changes in a systematic fashion by varying the between- and within-row spacings with all the rows parallel to one another.

Parameter

Parameter refers to the characteristics of a population in some respect and also the numerical quantity that specifies a population in respect to some characteristic.

Parasitism

A relationship between two living beings, where one has a negative effect on the other organisms.

Parenchyma

The thin-walled plant cells with a protoplast (the basic living components) whose function is food storage, photosynthesis (with plastids), wound healing and formation of adventitious structures.

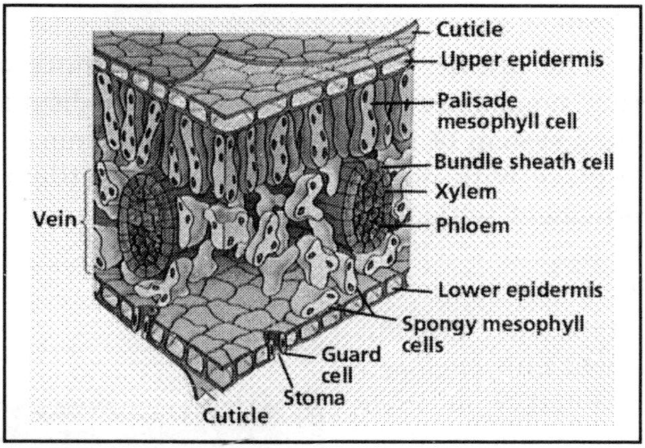

Parent Material

Refers to the horizon of weathered rock or partly weathered soil material from which the soil is formed.

Parthenocarpy

The process of formation of fruit without fertilization and without fertile seeds is called as parthenocarpy. Eg. Banana.

Parthenogenesis

The development of an organism from a sex cell without fertilization is called as parthenogenesis.

Particle Density (DP)

It is the mass per unit volume of soil particles and does not include the porespace. The value is expressed as g per cm^3 or in SI units as Mg per cubic meter, Mg m^{-3}.

Particle-Size Analysis

Determination of the various amounts of the different separates in a soil sample, usually by sedimentation, sieving, micrometry or a combinationof these methods.

Pastoralism

The term applies to a way of life for more or less wandering, animal-keeping peoples but nomadism strictly applies to a system of using land by moving animals from place to place with no permanent human settlement or semi-permanent settlement exists.

Pasture

A type of grazing management unit enclosed and separated from other areas by fencing or other barriers and devoted to the production of forage for harvest primarily by grazing.

Patch Cut

A clearcut on a small area is called as patch cut.

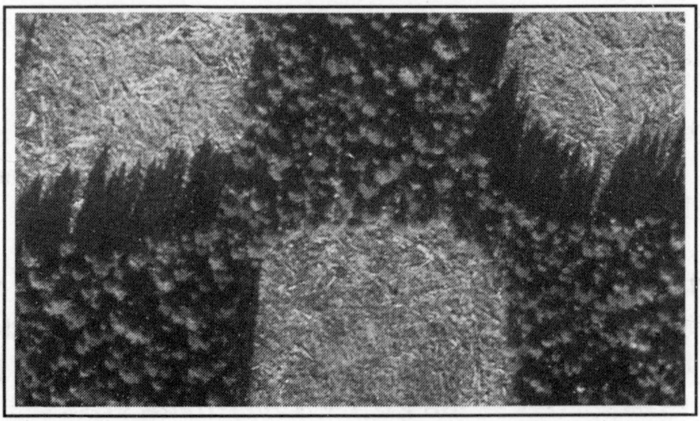

Pathogen

An agent capable of producing disease is called pathogen.

Pattern Analysis

A Methodology for analysing and displaying the underlying associations and relationships between sets of attributes is pattern analysis. It also refers to assessment of the regularity of an agroecosystem's structure and function, based on established data categories.

PC

A measure of the electrical conductivity of a soil for determination of ions in solution. The electrical conductivity of an aqueous suspension of soil is measured and expressed by the specific conductivity; that is, it uses the reciprocal ohm (or mho or millimho). pC is defined as the co-logarithm of the specific conductivity of soil extracts or suspensions in millimho cm–1.

Peat

Peat is an unconsolidated soil material that consists of undecomposed or slightly decomposed organic matter accumulated under conditions of excessive moisture.

Ped

A natural unit of soil structure, such as an aggregate, crumb, granule.

Pedalfer

A group of fully leached soils is called as pedalfer.

Pedigree Breeding

A system of breeding in which individual plants are selected in the segregating generations from a cross on the basis of their desirability, judged individually and from their pedigree record in self-pollinated crops.

Pedocal

A group of incompletely leached soils with free drainage is called pedocal.

Pedology

It refers to the frequency and duration of the periods when the soil is free of saturation or partial saturation. A measurable characteristic (including rapidity and extent) but generally assessed from profile morphology, e.g., greying and colour, and landform.

Pedon

The smallest volume that can be called 'a soil'. It has three dimensions; it extends downwards to the depth of plant roots or to the lower limit of the genetic soil horizons, and its lateral cross section is roughly hexagonal and ranges from 1 to 10 m in size depending on the variability in the horizon.

Penetrability

With soils, the ease with which a probe can be pushed in. May be expressed in units of distance, speed, force or work, depending on the type of penetrometer used.

Perched Watertable

The presence of water in a soil horizon as a result of poor internal drainage, topographic position or an impermeable soil layer. May be permanent or may occur during seasons of high precipitation.

Percolation

It is the downward movement of water through the soil in a saturated or nearly saturated soil.

Perennial

Perennial is a plant that does not die after flowering and lives from year to year.

Periphysis

Periphysis is the transmission to the ramet of effects caused by the environment of the ortet that can be reversible.

Permaculture

Permaculture is the design and maintenance of sustainable, ecologically favourable, energy efficient agricultural and horticultural systems. The concept includes not only agroforestry but the integration of organic farming principles and intermediate technology, the use of renewable resources and recycling, the exploitation of biodiversity, conservation and habitat protection, as well as social and institutional well-being.

Permanent Wilting Point

Permanent wilting point is the amount of water held in the soil when plants remain wilted even though when, for a short period, the aerial parts are kept in a humid atmosphere and usually taken as 15 atmospheres of tension.

Pest

In the general sense, an insect pest, fungal or viral pathogen, weed, or avian or mammalian pest (birds, rodents).

Pesticide

A substance, inorganic or organic, that kills or suppresses the growth or development of pests (insect pests, fungal pathogens, weeds).

PF

The logarithm of the free energy (expressed as a pressure or head in J kg–1) used to express the intensity with which water is held at any particular water content.

Ph

A numerical measure of the acidity, or hydrogen ion activity, of a soil. The neutral point is pH 7.0. All pH values below 7.0 are acid and all above are alkaline.

Phanerophytes

Self-supporting woody plants or herbaceous evergreen perennials that grow taller than 25–50 cm, or whose shoots do not die back periodically to that height limit are called phanerophytes.

Phareatic Surface

The upper surface or zone of saturation is called as phareatic surface.

Phase Change

Change from one phase (state) of plant development to another is called phase change. For example, from juvenile to mature phases.

Phenology

Phenology is the study of the timing of periodic phenomena such as flowering, growth initiation, growth cessation, especially as related to seasonal changes in temperature, photoperiod etc.

Phenophase

In phenology, the particular stage in development reached by a plant (organism), for example, germination, the juvenile phase (of a tree), anthesis (flowering).

Phenotype

The external feature(s) of an organism produced by the interaction of the genes of the organism with the environment is called phenotype. Appearance of an individual as contrasted with its genetic makeup or genotype. Also used to designate a group of individuals with similar appearance but not necessarily identical genotypes.

Phenotypic Plasticity

The capacity for phenotypic change shown by an individual genotype due to environmental influences is called phenotypic plasticity.

Phenotypic Selection

The process of selecting or choosing trees on the basis of their phenotypic performance.

Phloem

A compound vascular tissue in plants composed of sieve tubes, companion cells, fibres and parenchyma cells is called phloem.

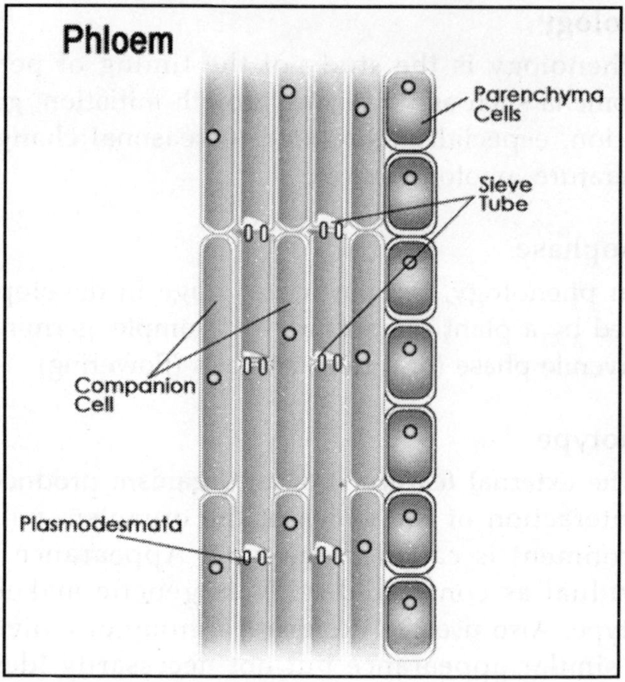

Photoperiod

It refers to a cycle of light and dark intervals, usually with a period of 24 hours.

Photoperiodism

The control of both reproductive and vegetative development through the imposition of appropriate sequences of light and dark periods is called photoperiodism.

Photorespiration

The light-dependent uptake of oxygen and release of newly fixed carbon dioxide from the substrate P-glycolate or glycolate which is derived from the photosynthetic carbon reduction cycle.

Photosynthesis

The process of conversion of light energy to chemical energy through a light (photon) capture by green plants and the conversion of atmospheric carbon dioxide to organic compounds.

Photosynthetically Active Radiation

In the electromagnetic spectrum, from 0.4 to 0.7 nanometers wavelength. Equals approximately 45% of the total shortwave radiation received at the earth's surface from the sun (abbreviated as PAR).

Phototropism

The growth (movement) of a plant organ as a response to the stimulus of light.

Phreatophyte

A plant drawing its water from subsurface sources, growing mainly along stream courses or where its roots reach the capillary fringe is called phreatophyte.

Phyllosphere

The space immediately adjacent to, and influenced by, a plant leaf is called as phyllosphere.

Phyllotaxy

The arrangement of leaves on a stem is called phyllotaxy.

Phylogeny

It refers to the racial history and evolution of a group of related organisms.

Physical Determinant

One of the important attributes of climate, water and land, such as rainfall, topography and hydrology, that influence configuration and performance of cropping patterns is called as physical determinant.

Physical Property

One of the characteristics, processes or reactions of a soil that are caused by physical terms or equations. Examples of physical properties are bulk density, hydraulic conductivity, pore-size distribution, porosity, water-holding capacity.

Phytohormone

A natural substance produced in small amounts that modulates growth or other plant developmental processes.

Phytosanitary Certificate

A written statement about the health of plant parts made when they are to be exported or imported. Usually in an internationally agreed form indicating field inspection, seed treatments, and so on.

Phytotoxic

Toxic (damaging) to at least some plants.

Pinching

It is the process of the breaking off the terminal growing point, allowing the axillary buds to start growing.

Pioneer

A tree species that has a high light requirement for seed germination and seedling establishment and is adapted to grow well in disturbed sites (forest gaps).

Pitting

The practice of planting young trees into prepared holes.

Plagiotropic Branching

A type of branching where stems grow only horizontally.

Plagiotropy

The gravitational response that produces an oblique or horizontal axis and results in dorsiventral symmetry and horizontal orientation.

Plant Development

It refers to the whole series of changes that an organism goes through during its life cycle involving growth and differentiation, but it can refer to individual organs, tissues or cells.

Plant Growth Regulator

An organic compound which may be a natural or synthetic substance that acts endogenously at very low concentrations to modulate plant development.

Plant Nursery

A specially prepared site for germinating seeds and looking after seedlings and young plants under conditions favourable for their growth and development. Facilities for vegetative propagation are also often present.

Plant Nutrient

An element essential to plant growth is called plant nutrient.

Plant Population

The term synonymous to plant density that refers to number of plants per unit area.

Plant Succession

An orderly process of change in the composition of a plant community until vegetation stabilizes and is in balance with its environment.

Planting Out

The procedure of moving and planting young plants that have been raised in a nursery to the site where they are to be grown next.

Plastochron

The time interval between two successive similar events, as in leaf development. Often used in a descriptive sense for its morphological results.

Plot

It refers to a group of trees, all from the same entry like family, clone, provenance etc that are planted together.

Plot and Contiguous

In a field study, a group of trees, all from the same entry such as family, clone, provenance that are planted together. Eg. A five-tree row-plot is probably the most common design for forest genetics experiments.

Plot and Non-Contiguous

In a field study, a group of trees, all from the same entry (family, clone, provenance . . .) planted in the same block, but not planted together which allows tests to be thinned systematically and maintain uniform spacing.

Pole

In forestry, a still-young tree, from the time its lower branches begin to die, up to the time when the rate of height growth begins to slow down and crown expansion becomes marked.

Pole Stand

It refers to a stand of trees whose average dbh is between 4 and 10 inches.

Pole Timber

The trees having 4 to 10 inches dbh.

Pollarding

The practice of cutting back in more or less systematic fashion the crown of a tree but leaving a main trunk to 1.5 m or so, with the object of harvesting small wood and browse, of producing regrowth beyond the reach of animals or of reducing the shade cast by the crown.

Pollination

The deposition of pollen on the receptive part of the female flower is called pollination.

Pollinator

Any plant that pollinates another. In horticulture, a variety or cultivar included where the main fruit tree variety is self-incompatible, so as to ensure cross pollination and proper fertilization so that fruit production occurs.

Polycross Test

It is a progeny test to assess general combining ability from crosses among selected parents where a mixture of pollen is artificially applied to each female parent, or the selections naturally and randomly open-pollinate themselves so that the identity is maintained for seed parents alone.

Polyploid

An individual or a cell having three (triploid), four (tetraploid), five (pentaploid), or more, complete sets of chromosomes instead of two as in diploids.

Population

Genetically refers to a group of similar individuals related by descent and so delimited in range by environment or endogenous factors as to be considered a unit. In cross-bred organisms the population is often defined as the interbreeding group.

Pore Space

Total space not occupied by soil particles is called as pore space.

Pore Size Distribution

The volume of the various sizes of pores in a soil, expressed as percentages of the bulk volume (soil plus total pore space) is called pore-size distribution.

Porosity

The volume of bulk soil not occupied by solid particles is called porosity.

Porous Wood

The wood with pores or vessels originating from a continuous cambium layer that is featured by nearly all hardwood species.

Post-Emergence

It refers to the period after the seedling has emerged from the soil.

Post-Emergent Application

The practice of applying a herbicide after the crop has emerged from the soil. In perennial crops it denotes any procedure carried out after the emergence of weeds.

Practice

In agroforestry, a particular use of land involving woody & non-woody plants in some spatial (simultaneous) or temporal (sequential) arrangement. For example, hedgerow, intercropping, homegardens, shifting cultivation.

Prairie

Nearly level or rolling grassland, originally treeless, and usually characterized by fertile soil.

Precision

Of an estimate, it is the closeness of agreement to be expected between a succession of independent estimates formed by a repetition of the sampling procedure.

Preclone

Any group of connected shoot-root units formed by a single plant and which has the potential to form a clone by the death or fragmentation of connecting tissue.

Precommercial Operations

It refers to the practice of cutting in forest stands to remove wood too small to be marketed for improving species composition and increase the quality, growth, and vigor of remaining trees.

Precommercial Treatments

Forestry operations that require landowner investment, such as cleaning or weeding stands to remove trees that have little or no cash value.

Predation

It is a relationship between living beings and where the whole of one organisms is taken away.

Pretreatment

The process or method of treating seeds for speeding germination for difficult species. Eg. scarification or treatment with boiling water.

Pricking Out

The procedure of transplanting young seedlings from the seedbed, seedbox or container in which they were germinated to a relatively more spacious area or container so as to give them room to grow.

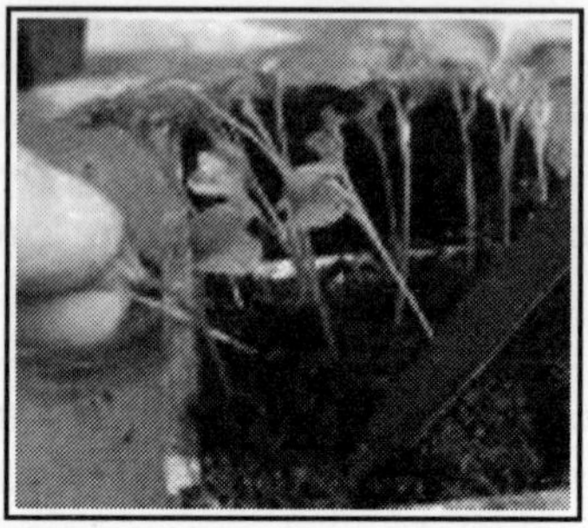

Primary Consumer

Organisms that subsists on plant material are called as primary consumers.

Primary Land Unit

A basic spatial unit suggested to embrace agriculture, horticulture, forestry and agroforestry. A homogeneous area in which a common species (or species mix) undergoes a common form of management (in agriculture, the plot; in forestry, the compartment).

Primary Producer

The organisms that produces organic material from solar energy & inorganic substance through photosynthesis are called primary producers.

Primary Production

Production of biomass by plants through the processes of photosynthesis and nutrient uptake. 'Net primary production' is this production less losses by respiration. It is often measured only as 'above-ground net primary production' and leaf fall should be accounted for.

Principal Components Analysis

A numerical method for ordination of data used either as a multi-variate (statistical) technique or purely for pattern seeking.

Principal Components

New variables derived by principal components analysis to describe a body of data. Each is a weighted sum of the 'raw' (as originally measured) variables, or of the centred or standardized variables.

Prizmatic

A type of soil structure with prism like aggregates with vertical axes longer than horizontal axes. Typically found in B horizons of Chernozemic and Solonetzic soils.

Procumbent

It refers to the condition of trailing or lying flat, but not rooting.

Production

The production refers to the output of an area of land.

Productivity

A measure of efficiency relating output of a product to the use of a resource (including time) is called as productivity.

Progeny

Progeny is an offspring of a particular plant or particular combination of one female and one male tree plant.

Progeny Test

Progeny test is the evaluation of parents by comparing the performance of their offspring evaluated under more controlled conditions than exist for the parent alone.

Prokaryote

An organism whose cells do not contain nuclei. Only the bacteria are prokaryotes.

Prolepsis

Prolepsis is the development of a lateral branch only after a period of dormancy as a lateral bud and the branch developed by such a process is called proleptic branch.

Propagate

To increase the number of a given plant type. Propagation can be by seed, root sucker, stool, stump, root, stem or leaf cutting, grafting, layering or by micropropagation.

Propagule

Propagule is a seedling, cutting or graft (especially when small) which has the potential for producing a new individual.

Prospectus

A document that describes the location of a property, indicates trees marked for cutting and states that the timber will be sold in accordance with a suitable contact. A prospectus includes the number of trees marked, their diameter classes, and a volume estimate for each species.

Protandry

The maturation of anthers before pistils is called protoandry. Eg. Maize.

Protective Plants

It includes the plants grown to protect crops, soils or land from adverse environmental factors.

Proteoid Root

It refers to one of the finely packed lateral roots which can form thick, subsurface mats. Thought to be microbially induced and found in some trees and shrubs of the Proteaceae. Eg. *Grevillea robusta.*

Protogyny

The maturation of pistils before anthers is called protogyny.

Protozoa

Protozoa are single-celled eucaryotic organisms. Eg. amoeba.

Provenance Test

A test comparing trees grown from seed or cuttings collected in many parts of a species range.

Provenance

The original geographic source of seed or propagules or the place in which any stand of trees is growing. The stand may be indigenous or non-indigenous.

Pruning

Pruning is the process of cutting back growth of plants, including roots, but more particularly, side branches of trees, or the sides and tops of hedges.

Psychrometer

An instrument used for measuring the water vapour content of the atmosphere; a type of hygrometer. It usually consists of two thermometers with wet and dry bulbs.

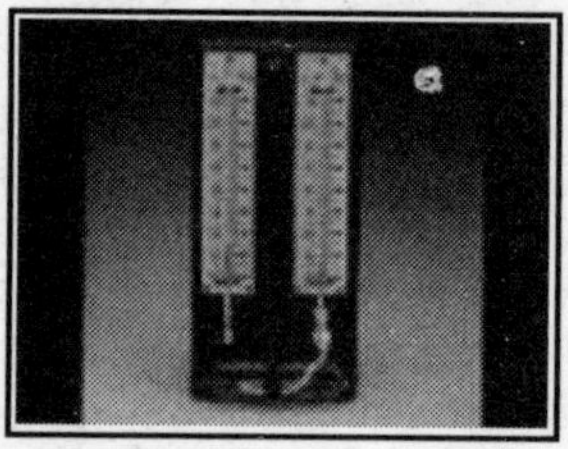

Psyllid

A small, hemipteran insect pest that sucks plant sap. The psyllid *Heteropsylla cubana* has spread around the world from its Central American home to devastate *Leucaena leucocephala* elswhere.

Puddled Soil

A dense soil often dominated by a single-particle structure, almost impervious to air and water. This condition results from handling a soil when it is in a wet, plastic condition, so that when dried, it becomes hard and cloddy.

Puddling

The practice of destruction of soil structure, reducing porosity and permeability. A dense soil of massive structure, almost impervious to air and water, occurs when wet, plastic soil is worked too much or is trampled by animals.

Pulaski

Firefighters use this tool, which combines a single-bitted axe blade for chopping with a narrow blade for trenching, to clear vegetation when constructing a fireline.

Pulpwood

The wood suitable for use in the manufacture of paper.

Pulses

It includes protein rich, non-oil-producing, cultivated legumes grown for their edible seeds.

Pure Line

A homogenous progeny obtained by successive self-fertilizations is called pure line.

—xxx—

Q

Quadrat

A small, clearly demarcated sample area of known size on which observations are made or it refers to an ecological sampling unit of any kind.

Quantitative Inheritance

Inheritance of characters that vary continuously and are controlled by three to many genes.

—xxx—

R

Race

A genetic subdivision of a species, with more or less distinct characteristics when grown in a particular environment. Sometimes used to denote a portion of a **cline.** Differences among races may or may not be adaptive; differences among ecotypes are presumed to be adaptive. Hence, 'race' is a more general term than ecotype or it refers to a genetically and often geographically distinct mating group within a species; also a group of pathogens that infect a given set of plant varieties.

Radial (surface)

A horizontal surface or plane extending wholly or in part from the pith to the bark of a tree bole.

Radicle

Radicle is the rudimentary root of the embryo.

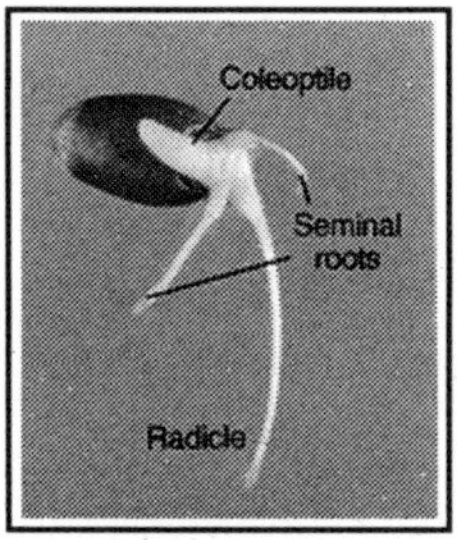

Rainfall Distribution

It is termed as the quantity of rain falling at different times of the year. The rain may fall over one season only ('monomodal') in tropics but near the equator it falls over two seasons and called as bimodal.

Rainfall Intensity

The rate at which rain is occurring at any given instant, usually expressed in millimetres per hour. 'Maximum rainfall intensity' refers to the rate during any time when the rate is both maximum and uniform. Usually one or more standard time periods, such as 5 or 10 minutes, are specified, and then the term refers to an average rate calculated from the amount that occurred during any such time period, although the rate was not neccessarily uniform or it refers to the amount of rain received per unit of time. Convective tropical storms often have a 'leading edge rate', that is, rainfall at the onset of the storm is greatest; it may be 150–200 mm h–1 or more for a short period.

Rainfed Farming

The term refers to the system of growing crops or animals under conditions of natural rainfall. Water may be stored in a crop field by bunding, as in lowland rainfed rice, but no water is available from permanent water storage areas.

Rainforest

A forest that grows in a region of heavy annual precipitation is called rain forest and it includes both tropical and temperate rainforests.

Ramet

An individual member of a clone, descended through vegetative propagation from the ortet or a unit of a plant with a set of roots and leaves. Ramets are connected by rhizomes or stolons.

Ramicorns

Abnormally large branches that project at sharp acute angles from the bole and are persistent (often associated with previous weevil attack) is called ramicorms.

Randomness

It refers to the condition of being unselected i.e. entirely by chance.

Range

A land which produces primarily native forage that is suitable for grazing by livestock but not for cultivation, especially in arid, semi-arid or forested regions. It also refers to the geographical and altitudinal limits within which a taxon occurs. But in case of statistics, the limits of magnitude (of a set of data) denotes range.

Rangeland

The land suitable for grazing by domestic livestock that includes vegetation like native grasses, grasslike plants, forbs, shrubs.

Ranger

An administrative officer in charge of a unit of forest or other land, usually a subdivision of a public forest or park. The various classifications include forest ranger, district ranger, park ranger, county ranger.

Rapid Rural Appraisal

It is a methodology in which a multidisciplinary team of researchers apply social science tools for the quick collection of primary data during one or several visits. They transform the primary data into secondary data for assessment and analysis.

Rate Variable

It is a variable which controls the rate at which various responses are generated by various changes in the system's state. Rate variables represent the coefficients of the governing equations describing the dependence of endogenous variables on their controlling state variables.

Ratoon

A sprouting or regrowth that arise from buds or young tillers; crops originating from ratoon growth.

Ratoon Cropping

The practice of cultivation of an additional crop from the regrowth or stubble of a main crop after its harvest, by avoiding replanting is called ratoon cropping. Eg. sugarcane, sorghum and rice.

Rattan

A climbing palm (mainly *Calamus* spp and others in the lepidocaryoid line). Used for making baskets and furniture.

Ray

In wood anatomy, a ribbon-shaped strand of tissue formed by the cambium and extending in a radial direction across the grain in hardwoods.

Recalcitrant Seed

The seeds which have relatively short viability and cannot be stored in a dry condition or at low (subzero) temperatures due to chilling damage. Examples are seeds of cocoa and rubber.

Recessive

The member of a pair of alleles that is not expressed when the other (dominant) member occupies the homologous (corresponding) chromosome is called recessive.

Reciprocal Crosses

The crosses in which the sources of male and female gametes are reversed is called as reciprocal crosses.

Recombination

Recombination is the formation of new combinations of genes through crossovers and independent segregation of chromosomes at meiosis by exchange of chromosome segments (the division of sexual cells).

Recommendation Domain

It is a more or less homogeneous group of farmers with similar circumstances so that similar recommendations can be made.

Recovery Rate

The rate of conversion of roundwood to timber products is called as recovery rate.

Recurrent Selection

A method of breeding designed to concentrate favourable genes scattered among a number of individuals by selecting in each generation from among progeny produced by matings of all the selected individuals (or their selfed progeny) of the previous generation.

Reforestation

The term refers to the replacing forests after felling.

Regeneration Class

Regeneration class refers to the area, and the young trees in the area, being managed during the regeneration interval in the shelterwood silvicultural system. Here old and young trees occupy the same area, where the old trees protect the young trees.

Regeneration Cut

Any removal of trees intended to assist regeneration already present or to make regeneration possible is called regeneration cut.

Regeneration Initiation

It refers to the year in which the new crop is deemed to be started at an acceptable stocking level, whether by planting, natural or artificial seeding, or by vegetative means.

Regeneration Interval

The period between the seed cutting and the final cutting on a particular area under one of the shelterwood systems.

Regeneration Period

It applies to the time between the initial regeneration cutting and the successful re-establishment of a stand by natural or artificial means.

Regional System

It is a complex, large scale land unit of utilization that produces and transforms primary products and involves a large service sector.

Registered Seed

The progeny of foundation seed that is normally grown for producing certified seed is called registered seed.

Regular Uneven Aged Structure (balanced)

A stand in which three or more distinct age classes occupy approximately equal areas and provide a balanced distribution of diameter classes.

Reiteration

Development of shoots outside the normal expression of the architectural model of a tree, as a specialized environmental response, but repeating the basic model is called reiteration.

Reiterative Complex

It refers to the complex of meristems that is produced by reiteration.

Relational Diagram

A diagrammatic representation used to show the interrelationships of components and processes in a system is called relational diagram.

Relative Growth Rate

At a given instant, the increase in plant material per unit of plant material initially present. Expressed over a period (for example, one week) as a mean relative growth rate in g g–1 day–1.

Relative growth rate = net assimilation rate x leaf area ratio.

Relative Humidity

The dimensionless ratio of the actual vapour pressure of the air to the saturation vapour pressure, which changes with temperature.

Relative Thinning Intensity

It is defined as the periodic (annual) yield of a stand from thinnings, expressed as a percentage of its periodic annual increment.

Relay Cropping

The system of planting crops between plants or rows of an already established crop during the growing period of the first planted crop(s). One form of overlapping cropping or it refers to the growing of two or more crops on the same field with the planting of the second crop after the first one has already completed development.

Release Cutting

A cutting operation which is carried out to release the forest crop (seedlings or saplings) from competition with undesirable tree species of similar size (termed a cleaning) or larger and overtopping tree species (termed a liberation cut).

Release

It refers to the practice of freeing a tree or group of trees from more immediate competition by cutting or otherwise eliminating growth that is overtopping or closely surrounding them.

Replication

Refers to the number of times a treatment is repeated in space or time. To halve the standard error, four times the degree of replication is needed, so extraneous variability must also be identified and dealt with or applying a treatment or set of treatments more than once to increase the precision of comparisons and to provide an assessment of the variability among experimental units treated alike.

Representative Sample

A sample whose properties accurately reflect those of the population that is large enough to allow statistical treatment and conclusions about the population as a whole. Samples can be 'simple random samples' (a type of sampling unit), 'multistaged random samples' (a series of sampling stages, each with its sampling unit), 'stratified random samples' (where the sampling units are first assigned to strata and these are then randomly sampled) and 'stratified multistage random samples'.

Reproduction

Reproduction is the process by which a forest is renewed. **A. Artificial.** Renewal by direct sowing or planting (syn. reforestation). **B. Natural.** Renewal by self sown seeds, sprouts, rhizomes, etc. (syn. regeneration). 2. Seedings or saplings of any origin (syn. young growth).

Reproduction Methods

1. Clearcutting. Removal of the entire forest in one cut that perpetuates even-aged stands. 2. seed-tree. Removal of the mature timber in one cut, except for a small number of seed trees (1 -7 trees/acre); called a "group cutting" when the seed trees are left in groups, a "reserve cutting" when specifically selected seed trees are left for growth, as well as to furnish seed. 3. selection. Removal of mature timber, usually the oldest or largest trees, either as single scattered trees or in small groups at relatively short intervals, commonly 5 to 20 years, repeated indefinitely which encourages a continuous establishment of natural reproduction, and an uneven-aged stand is maintained. Also called "thinning from above." 4. Shelterwood. The removal of the mature timber in a series of cuttings, which extend over a period of years. Usually equal to not more than one-quarter (often not more than one-tenth) of the time required to grow the crop. The establishment of natural reproduction under the partial shelter of seed trees is encouraged, but sometimes these areas must be artificially regenerated. 5. Coppice. The forest regeneration by sprouting (vegetative reproduction) from stumps or roots.

Research

Research is an investigation directed to the discovery of some fact by the careful study of a subject. A course of critical or scientific enquiry. The act of researching, which, in its broadest sense, involves both inductive (algorithmic) reasoning and heuristic convergence.

Research Strategy

The allocation of research resources to specific activities to maximize the efficiency and effectiveness of research according to certain specified goals (such as improving the sustainability and availability of food to all sectors of the population).

Residual Herbicide

A herbicide applied to the soil that remains active for at least several weeks, but sometimes many years.

Residual Stand

Trees, often of saw log size, left in a stand after thinning to grow until the next harvest. Also called "reserve stand" or "leave trees."

Resins

A class of inflammable vegetable substances that have no definite shape, secreted by certain plants or trees; characterizes the wood of many coniferous species.

Residual Variation

Residual variation is the part of the total variation in data that cannot be explained by differences between treatments or blocks or by the effects of other controllable or measurable variables.

Resilience

It is defined as the ability of a population to persist in a given environment despite disturbance or reduced population size. The resilience of a population is based upon the ability of individuals within the population to survive (fitness) and reproduce (fecundity) in a changed environment and the genetic variability of the population which allows the production of new genotypes.

Resin

A thick, sticky liquid that comes out of certain trees and later becomes hard. *Acacia senegal* produces one of the more valuable resins ('gum arabic').

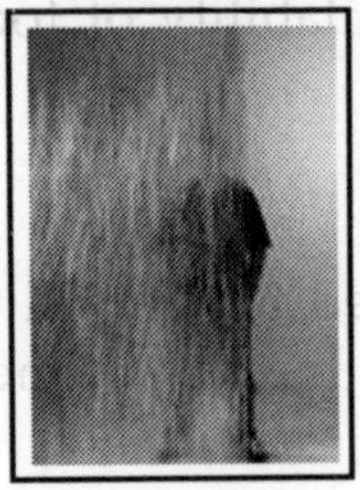

Resistance

It is defined as the ability of an organism to exclude or overcome, completely or in some degree, the effect of a pathogen or other damaging factor. It may be horizontal or vertical resistance.

Resource Capture Efficiency

It is the measure of the ability of a plant, or more often a plant community, to transform captured resources (light, water, nutrients) to dry matter or yield.

Resource Capture

The acquisition by plants of environmental resources, that is, the processes by which plants obtain light, water and nutrients.

Respiration

Respiration is the processes by which tissues and organisms exchange gases with their environment.

Response

The measured reaction to some treatment is called response. Eg. the effect of pruning the basal area of trees or of thinning on the diameters of branches of the trees remaining.

Rhizobium

Type of bacterium that has the capacity of both invading the roots of certain species of the Leguminosae and fixing atmospheric nitrogen, which is subsequently used by the host plant.

Rhizome

A horizontal stem growing beneath the soil surface, capable of producing adventitious roots and with nodes having scale leaves subtending axillary buds that may form either rhizome branches or upright shoots.

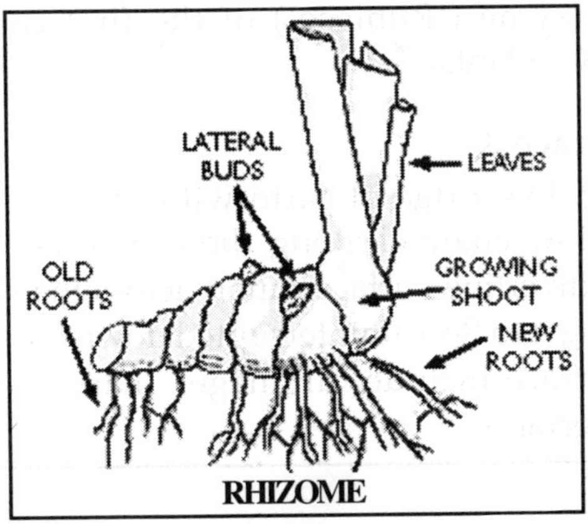

Rhizophore

Refers to a leaf-bearing root or a leafless stem with roots.

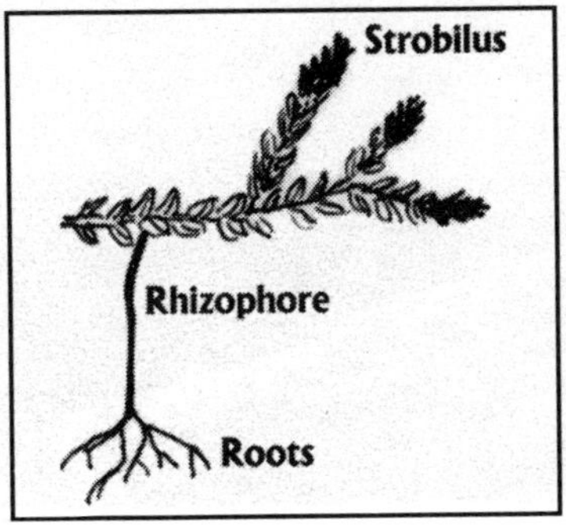

Rhizosphere

Rhizosphere is the soil close to plant roots where there is usually an abundant and specific microbiological population.

Rhythmic Growth

The growth of an axis determined by a rhythm that results in periodic shoot extension alternating with dormancy is called as episodic growth.

Ribulose Biphosphate

The enzyme ('Rubisco') of the first carboxylation process in C_3 plants.

Ridge Terrace

A long, low ridge of earth with gently sloping sides and a shallow channel along the upper side, to control erosion by diverting surface runoff across the slope instead of permitting it to flow uninterrupted down the slope. Types of ridge terrace include 'drainage', 'narrow-based', and 'Nichol's' terraces.

Rill Erosion

The formation or occurrence of many small channels a few centimetres deep due to poor cultivation is called as rill erosion.

Rill

Rill is a small, intermittent water course with steep sides which is usually only a few centimetres deep.

Ring-Barking

It is a method of killing trees by removing a ring (strip) of bark right down to, and including, the cambium near the bottom of the trunk.

Ring-Porous Wood

The wood in which the pores of one part of a growth ring are seen in distinct contrast in size or number (or both) with those of the other part.

Riverine

The creation of different environment along the permanent and semi-permanent streams because of the increase in soil moisture is termed a riverine environment.

Roguing

The systematic removal of undesired individuals in a population is called rouging.

Root Climber

Root climber is a plant that raises itself up by means of adventitious roots, which adhere to the plants that support it.

Root collar

The transition zone between stem and root. Usually recognizable in trees and seedlings by the presence of a slight swelling.

Root Hair

The unicellular, hairlike growth from a root involved in water and nutrient uptake.

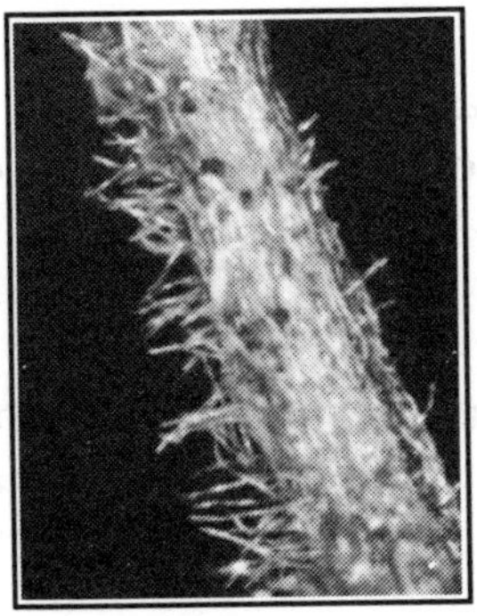

Root Sucker

Root sucker is a shoot arising from an adventitious bud in a root that can be cut away and planted at another site. Eg. Banana and datepalm. Other trees will produce root suckers if their roots are cut at some distance from the tree and left exposed (for example, *Ocotea usambarensis* and *Melia volkensii* can be propagated in this fashion).

Root to Shoot Ratio

It refers to the ratio of oven dry root weight to oven dry shoot weight in seedlings.

Root Zone

Layer of the soil interwoven by plant roots. Also called 'rooting zone'.

Roots

The below-ground tree or plant parts that provide physical support, absorb water and nutrients from the soil, and store food produced by photosynthesis.

Rotation

The period of years required to establish and grow a timber crop to a specified condition of maturity, when it may be harvested and a new tree crop started.

Rotation Age

It refers to the age at which a stand is considered ready for harvesting under an adopted plan of management.

Rootstock

A root bearing plant or plant part, generally a stem or root, onto which another plant part (the scion) is grafted. It also refers to the collective roots in a stand, capable of sprouting.

Rot

Wood in a state of decay.

Rotational Cropping

The repetitive cultivation of an ordered succession of crops, or crops and fallow, on the same land. One cycle often takes several years to complete.

Rotational Grazing

Grazing systems in which the pasture is subdivided into a number of enclosures, with at least one more of these than there are groups of animals. The practice of submitting a grazing area to a regular sequence of grazing alternating with rest.

Roundwood

The timber as harvested, before conversion is called round wood.

Row Intercropping

The practice of growing two or more crops simultaneously in which, one or more crops are planted in rows is called row intercropping.

R-Selection

The selective conditions which favour high reproductive rates at the expense of individual vegetative longevity is called R selection.

Runner

A slender horizontal stem with elongated internodes that roots at nodes that touch the ground.

Runoff

The water, usually from rain, that fails to infiltrate the soil surface and passes out of the plot or field is called as runoff.

—xxx—

S

Sale Area

The land area which contains the trees that are to be sold or the area that will be affected by the harvesting operations.

Sale Boundary

The term refers to the area within which the harvesting operation is to be conducted which should be clearly marked by the seller.

Sale Log

The sale of sawlogs or sawtimber site trees.

Sale Lump Sum

A specified volume of standing trees sold for a cash price prior to cutting. The trees may be marked or tallied or the whole tract cruised and the volume computed.

Sale Mill

Sale paying the landowner for the volume of lumber produced at the sawmill.

Saline Soil

Saline soil is a non-alkali soil that contains enough salts to interfere with the growth of most crop plants. Conductivity of the saturated extract is greater than 4 mmhos cm –1, exchangeable sodium percentage is > 15, and the pH is usually > 8.5. saline soil containing high amounts of the sulphates and chlorides of sodium and calcium. Saline soils occur in hot climates, have a typically uneven surface, show little change down the profile and are low in humus. If the sodium salt is carbonate then the reaction can be around pH 9.0. Highly saline soils can contain as much as 1% of salts in the topsoil (that is, up to 250 t ha–1 in the top 120 cm).

Salinity

The concentration of dissolved salts in water (g kg–1) when the organic matter has been oxidized, the carbonate converted into oxides, and the bromine and iodine converted into chlorine.

Salvage Cut

The harvesting of trees that have been or are soon to be damaged beyond recovery (by insects, disease, or the environment) in order to utilize the timber before further decline occurs.

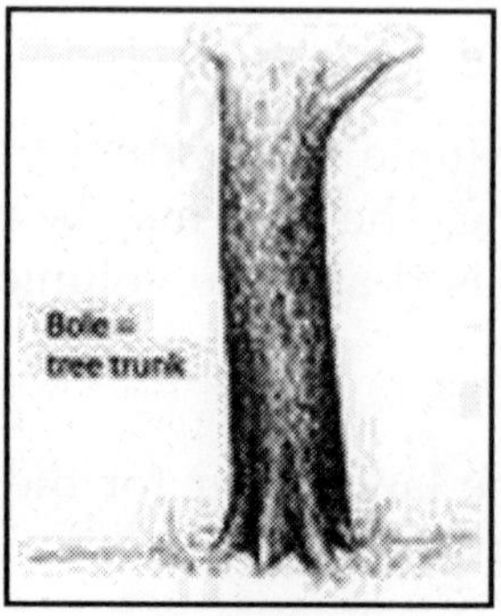

Salvage Cut

The practice of harvesting dead trees or those in danger of being killed (by insects, disease, flooding, or previous fire) to save their economic value and promote forest health.

Salvage

It refers to a practice of cutting trees that are dead or in poor condition but can still yield a forest product.

Sample

A part of a population, consisting of one or more units selected and examined as representative of the whole. Because of the variability in biological populations, little information is to be gained by measuring just one individual, so a number are taken to form a 'sample'. The larger this is, the more closely its characteristics (such as the mean) represent the **population.**

Sample tree

A representative or average-sized tree, chosen for detailed measurement of condition, size, growth, or quality is called as sample tree.

Sampling

Sampling refers to taking detailed measurements of selected small parts of a forest in order to gain information about the whole forest.

Sampling Unit

Sampling unit refers to an individual **plot** or **quadrat.** A collection of many such units, each of which is a different small fragment of the community under study, constitutes a **sample.**

Sand

Particles between 2.00 and 0.05 mm diameter, or one of several separates such as coarse or medium sand; a soil textural class or it refers to a mineral or rock fragments that range in diameter from 2.00 to 0.02 mm in the international system, or 2.00 to 0.05 mm in the USDA system.

Sanitation Cut

Sanitation cut is the removal of trees to improve stand health by stopping or slowing the spread of disease or insects.

Sanitation Measures

The removal of dead, damaged, or susceptible trees or their parts, or of vegetation that serves as alternative host for crop-tree pathogens, to prevent or control the spread of pests or pathogens.

Sap

The liquid that rises from the roots of a tree. Sap consists of water and minerals. In the spring it also contains sugars that stimulate growth.

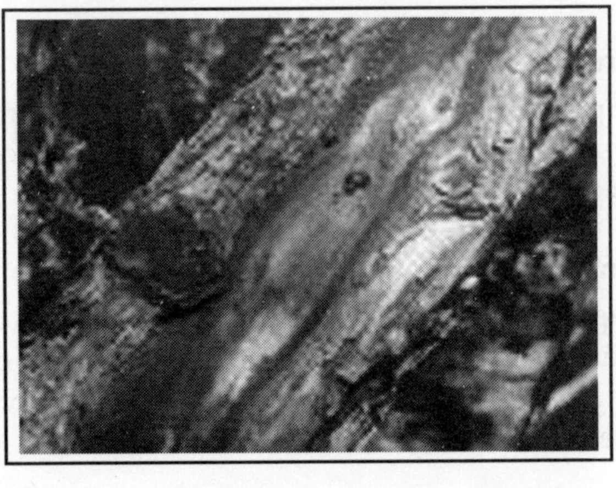

Sapling

Begins with the end of the seedling stage and ends when trees reach 10 cm diameter at breast height (dbh), the crowns are well elevated, and usually many of the lower branches have died or sapling refers to a loose term for a young tree no longer a seedling but not yet a **pole,** that is, a few metres high and 2–3 cm in diameter at breast height (dbh), growing vigorously and without dead bark or more than an occasional dead branch or a small tree, usually between 2 and 4 inches DBH is called sapling.

Sapling Stand

A stand of trees whose average dbh is between 1 and 4 inches.

Sapwood

The light colored wood which appears on the outer portion of a cross section of a tree and composed of dead cells; serves to conduct water and minerals to the crown. A minimum of 1 in of sapwood is required on all poles to ensure proper absorption of preservatives. Also termed "xylem."

Saskatchewan Forest Habitat Project

A partnership established in 1989 involving Wildlife Habitat Canada, Saskatchewan Wildlife Federation, The Federation of Saskatchewan Indian Nations, Weyerhaeuser Canada Saskatchewan Division, Saskatchewan Environment and Resource Management, Forestry Canada through the CanadaSaskatchewan Partnership Agreement in Forestry and Prince Albert National Park to cooperatively balance timber production and wildlife habitat objectives.

Saturation Vapour Pressure

The maximum possible partial pressure of water vapour in the air or atmosphere at a given temperature is called saturation vapour pressure.

Saturation Water Vapourpressure Deficit

The degree of 'dryness' of the atmosphere. The difference between the saturation vapour pressure at the existing temperature and the actual vapour pressure. Also termed 'saturation deficit'.

Savanna

A grassland with scattered trees, either as individuals or clumps. Often a transitional type between true grassland and forest.

Saw Timber

It refers to trees of a size and quality suitable for sawing into timber.

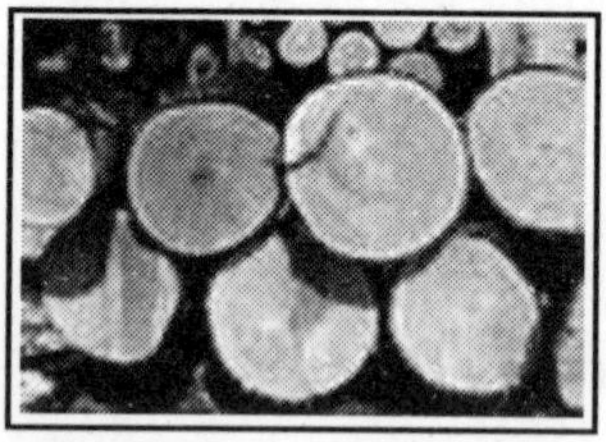

Saw Timber Stand

A stand of trees whose average dbh is greater than 11 inches.

Sawlog

A log designated to produce a sawn product usually at least 30 cm in diameter and the length of the logs to be cut will depend on the market requirement or a log large enough to be carved into lumber, usually at least 10 to 12 inches in diameter or it refers to a log which is large enough to be sawn into lumber.

Sawlog Tree

A tree at least 11 inches dbh and suitable for conversion to lumber. Sometimes, trees 11 to 14 inches dbh are called small sawlog trees, and trees larger than 18 inches dbh are called large sawlog trees.

Sawmill

A plant at which logs are sawed into salable products, including all the machinery and buildings necessary for the operation of the plant.

Sawtimber Stand

A group of trees with individual trees large enough to be sawed into lumber.

Sawtimber

Trees that will yield logs suitable in size and quality for the production of lumber or it refers to the trees that yield logs suitable in size and quality for the production of lumber.

Scale

The estimated sound volume of a log or group of logs in terms of a given log rule or formula that is used to estimate the sound volume of a log or group of logs.

Scale Stick

A flat stick, similar to a yardstick, calibrated so that the log volume can be read directly when the stick is placed on the small end of a log of known length.

Scale Ticket

A receipt provided by the processing facility that shows the volume of timber measured at the entry. Measurement may be by scale or by weight.

Scale and Log

It refers to the volume of roundwood products such as logs expressed in board feet, cubic feet, etc.

Scale and Mill

The term which is also called as mill tally refers to the volume of sawed wood products such as lumber.

Scalp

To physically remove the sod or surface layer of debris, to expose mineral soil for tree planting. scarify. To disturb the forest floor and top soil in preparation for natural regeneration or direct seeding or planting.

Scarification

Scarification refers to the abrasion of the seed coat (or fruit coat) by mechanical, chemical or physical (for example, dry heat) means to improve the germination of hard seeded species.

Scarify

The practice of disturbing the forest floor and top soil in preparation for natural regeneration or direct seeding or planting.

Schoolmarm

The Logger's slang for a tree with one or more trunks.

Scion

Scion is a twig, or portion of a twig, of one plant that is grafted or budded onto the root stock of another or it can also refers to any unrooted portion of a plant used for grafting or budding onto a rootstock. Shoots of woody plants from which scions are cut are called 'scionwood' or, if approprate, 'budwood'.

Scrambler

A plant that raises itself above other vegetation in the understorey with the aid of thorns or hooks is called as scrambler.

Screening

It is a preliminary comparison of a large number of treatments for evaluating and eliminating obvious undesirable treatments, as in 'wide range provenance screening trials'.

Scribner Log Rule

A method of estimating tree volume that disregards taper and underestimates the actual scribner volume produced at the mill.

Scrub

A vegetation type that is intermediate between forest, bushland and thicket that implies a poorly productive area.

Sealed-bid sale

A timber sale, usually offered through a consulting forester, in which buyers submit secret bids.

Seasonal Plant

A plant that flowers and completes its life cycle within the duration of a single wet dry season combination (in equatorial regions) is called as seasonal plant.

Seasoning

The process of drying (curing) lumber or other forms of wood to improve its properties: natural (air or underground drying) or artificial (kiln drying, electrical drying, oil drying, etc.).

Season

To dry lumber, either in the open or in a dry kiln.

Second growth

A second forest that develops after harvest of the original, natural forest. In the Pacific Northwest, these forests also are often called young growth stands.

Section

A unit of land area equal to 640 acres, 6,400 sq chains, I sq mile, or 80 chains on each side.

Second Growth

The forests that reproduce after removal of the original forest by cutting, fire or other cause is called second growth.

Secondary Production

The production of biomass (by animals, microorganisms or parasitic plants) through the use of primarily produced plant materials.

Secondary Succession

A succession initiated by an abiotic or biotic agent or agents after the ground has been cleared of its original vegetation is called secondary succession.

Sediment

A solid material, mineral or organic, in suspension in transport, or that has been removed from the original site by air, water, gravity, or similar agents.

Seed Bed

In natural regeneration, the soil or forest floor on which seed falls. In nursery practice, a prepared area over which seed is sown.

Seed Collection Area

A forest stand that exhibits good characteristics of growth, form, and vigour and that is not managed for cone production, but from which quality seed is collected, usually at time of harvest.

Seed Cutting

Removing trees in a mature stand so as to effect permanent opening of its canopy (if there was no preparatory cutting to do this) and so provide conditions for securing regeneration from the seed of trees retained for that purpose; the first of the shelterwood cuttings under a shelterwood system.

Seed Orchard

A plantation of trees, assumed or proven genetically to be superior, that has been isolated so as to reduce pollination from genetically inferior outside sources, and intensively managed to improve the genotype and produce frequent, abundant, and easily harvestable seed crops. A clonal seed orchard is established by setting out clones as grafts or cuttings; a seedling seed orchard is established from selected seedling progenies.

Seed Orchard and Clonal

A seed orchard established from vegetatively propagated trees, usually grafts. Usually clonal seed orchards are established with single tree plots with different ramets of each clone located as far apart as possible to reduce self pollination.

Seed Orchard and Seedling

A seed orchard established from seedlings (as opposed to grafts). Usually seedling seed orchards are established with multiple tree plots so that selection can take place first among families and then among the individuals within each family plot, reducing each plot to a single tree.

Seed Production Area

A forest stand identified as a good source of seed and in which individual trees are evaluated for desired characteristics. Unwanted trees and competing trees are removed to promote seed production. Seed is collected during the life of the stand or a stand or plantation not originally planted to produce seed, but rogued of inferior trees and treated in such a manner as to produce large quantities of seed. Usually the stand or plantation is from a known origin or selected seedlot.

Seed Source

The locality where a seed lot was collected. If the stand from which collections were made was exotic, the place where its seed originated is the original seed source.

Seed Spot

A prepared, limited space, e.g., a small, cultivated patch, within which (tree) seeds are sown.

Seed Transfer Guide

A set of rules for collecting seed and planting seedlings so that genotypes are not moved to microclimates or soils within which they will be mis-adapted. Usually seed transfer guides describe the maximum movement from the point of collection in miles east and west, north and south as well as feet in elevation.

Seed Trap

A device for catching the seeds falling on a small area of ground, from trees or shrubs. Used for determining the amount of seedfall and the time, period, rate, and distance of dissemination.

Seed Tree

A tree that produces seed or trees reserved in a harvest operation to supply seed.

Seed Tree Method

The removal of trees from the harvest area at one time except for a few scattered trees left to provide seed to establish a new forest stand.

Seed year

A year in which a given species produces (over a considerable area) a seed crop greatly in excess of the normal which is applied usually to trees of irregular or infrequent seed production.

Seed Zone

An area within which seed can be collected from any natural stand and planted in any new site without fear of mis adaptation.

Seedbed

In natural plant reproduction, the soil or forest floor on which seed falls; in nursery practice, a prepared area in which seed is sown.

Seeding

A reforestation method by sowing seed aerially or by hand which is often done immediately after harvest so that a new forest is started the next growing season.

Seedling

A small tree grown from seed. Usually the term is restricted to trees less than 2 in d.b.h. or **a** young tree, grown from seed, from the time of germination to the sapling stage, having a diameter at breast height equal to or less than 1 cm. A tree, usually defined as less than 5 cm in dbh, which has grown from a seed (in contrast to a sprout). **bare-root**: A seedling to be planted with its roots bare of soil. **container**: A seedling grown in a container and that is to be planted with roots still in its growth medium. **plug**: A seedling lifted from its container with roots and rooting medium left undisturbed.

Seeding Aerial

Broadcast seeding of seeds or seed pellets from aircraft. **broadcast**: The sowing of seeds more or less evenly over a whole area on which a forest stand is to be raised. **direct seeding**: The artificial systematic sowing of seeds by manual or mechanical means in an area on which a forest stand is to be raised. **drill**: The sowing of seeds in shallow furrows across a whole area on which a forest stand is to be raised. **natural seedling:** The dispersal of seed by natural agents from standing trees in proximity to a regenerating area or from slash scattered over that area. Seeds may be dispersed by wind, birds, mammals, gravity, or flowing water. **spot:** The sowing of seeds within small, cultivated, or otherwise prepared patches, many of which are distributed over a whole area on which a forest stand is to be raised.

Seedlot

A convenience term denoting a group of seeds, or their offspring, which will be considered as a unit in an experiment. Seedlot refers to an indefinite quantity of seed having uniform quality produced at a specific location and collected from a single crop.

Seed Tree Harvest Cut

A harvest cut on which all of the trees are removed from the harvest area at one time, except for a few scattered trees that are left to provide seed to establish a new forest stand.

Seed Tree Harvest

The felling of all the trees in an area except for a few desirable individuals that provide seed for the next forest.

Seed Tree Method

A method of regenerating a forest stand in which all trees are removed from the area except for a small number of seed bearing trees that are left singly or in small groups. The objective is to create an even aged stand.

Segregation

The process of separation of genes or chromosomes of maternal and paternal origin at meiosis (the division of sexual cells) the reduction division in gametic cells is called as segregation.

Selected Seed

The seed attested by a recognized certifying agency to be from progenitors previously selected as superior phenotypes for one or more characteristics but on which tests of inheritance are either pending or lacking.

Selection

Choosing individual trees or populations with desirable characters to obtain genetic improvement.

Selection Cutting

Annual or periodic cutting of trees chosen individually or by groups, in an uneven-aged stand, in order to recover the yield and develop a balanced uneven-aged stand structure, while providing the cultural measures required for tree growth and seedling establishment. The cuts are usually a mix of regeneration cuts and improvement cuts.

Selection Differential

The average phenotypic value of the selected individuals, expressed as a deviation from the population mean.

Selection Harvest

The harvest of all individual trees or small groups at regular intervals to maintain an uneven-aged forest. Selection harvests are used to manage species that do not need sunlight to survive.

Selection Index

Choosing parents on the basis of several desirable traits. The desirability of an individual, family or clone is judged according to an index value calculated by considering the heritability and importance of every trait.

Selection Method

A method of regenerating a forest stand and maintaining an uneven-aged structure by removing some trees in all size classes either singly or in small groups or strips.

Selection Pressure

Selection pressure is the strength of the tendency to eliminate undesirable genotypes or phenotypes and it is expressed in terms of a selection differential or as the portion of total trees that are selected.

Selection

Selection is the choosing of individuals or populations with desirable characteristics to obtain genetic improvement.

Selection System

The removal of mature trees, either as scattered individuals or in small groups, at relatively short intervals and repeated indefinitely. The objective is representation of all ages of trees throughout the stand to maximize growth of a quality product on a continuous basis.

Selection Thinning

Removal of dominant trees to benefit trees in lower crown classes.

Selection and Clonal

Choosing the best clones from a clonal test.

Selection and index

Choosing parents on the basis of a weighted score which combines economic value and heritability of several desirable traits.

Selection, mass (Phenotypic Selection)

Choosing trees on the basis of their phenotypes (appearance) alone.

Selection, recurrent

Choosing individuals out of families, making crosses, then progeny testing. The process is usually repeated several to many times. Most tree improvement programs are based on recurrent selection.

Selection, tandem

Selecting for two or more traits sequentially rather than simultaneously. The technique is useful when one trait can be evaluated early and a second trait only after a number of years, or when one trait is much more expensive to measure.

Selective Cutting

A system of cutting in which single trees, usually the largest, or small groups of such trees, are removed is called selective cutting.

Selective Herbicide

A herbicide that kills only certain groups of plants without injury to others. Eg.2,4 D, which kills broadleaf plants but not grasses.

Selective Thinning

Removing chosen trees, usually so as to give the maximum benefit to those remaining is called selective thinning.

Self Pollination

The natural or artificial pollination of a female flower with pollen from the same genotype.

Self

To place pollen from a male flower on a female flower on the same plant is called self.

Self-Incompatibility

Self incompatability is the inability to produce seed following self pollination where it is limited specifically to cases in which the inability is due to a pollen borne gene that prevents pollen tube growth on a stigma having the same gene.

Self-Mulching Soil

A soil with a high swelling potential in the surface layers so that it cracks into a granular mulch on drying is called self mulching soil.

Self-Sterility

The inability to produce seeds following self pollination is called self sterility.

Seller

The person offering the timber for sale, usually the owner, but may be a manager.

Senescence

The deteriorative changes that occurs in an organ or whole plant preceding death (organ senescence, whole plant senescence) which can also be 'sequential senescence' (leaves along a branch) and 'synchronous senescence' (all at the same time).

Sensitivity Analysis

It refers to the method of discovering by how much the estimates used in a model can vary without changing the final result.

Sequential Cropping

A pattern of growing more than one crop on the same piece of land with each seasonal crop component being grown during a different time of the year.

Sericulture

The culture of silkworms *(Bombyx mori)* for the purpose of production of silk is called sericulture.

Serotinous

Coming late; particularly applied to plant species or individuals that flower or fruit late in the season and to fruit or cones that remain on the tree without opening for one or more years (e.g., *Pinus contorta* and *Pinus banksiana),* but also to bud opening and leaf shedding.

Service

A beneficial attribute in agroforestry, (apart from products) brought about by a particular practice like shelter, soil fertility improvement, soil and water conservation.

Service Forester

A professional forester or Service foresters help private woodland owners develop and implement forest management plans.

Sessile

Sessile refers to the condition of a leaf, without a petiole (leaf stalk), that is, borne directly on the stem.

Severance Tax

A tax on forest products after they are cut. It is usually paid by the first processor.

Shade Intolerant

A tree relatively incapable of developing and growing normally in the shade of, and in competition with, other trees.

Shade Tolerance

The capacity of a tree or plant species to develop and grow in the shade of and in competition with other trees or plants.

Shade Bearing Tree

The tree species which will regenerate in shade so is often large seeded and is also known as 'climax species' is called as shade bearing tree.

Shade Tolerant

The capacity of a tree to survive and grow in the shade of, and in competition with, other trees.

Shake

A lengthwise separation of wood (usually caused by wind) that usually occurs between and parallel to the growth layers or a thin section split from a bolt of wood and used for roofing or siding.

Shatter

The falling off and consequent loss of potentially harvestable foliage or fruits.

Shear

In Christmas tree culture, it refers to shape and trim back the branches to make dense foliage and give tree a conical shape, but in felling it refers to a mechanical device that pinches trees off at the stump, natural (air or underground drying) or artificial.

Shearing

A method of harvest using mechanical shears or the shaping of a tree crown, particularly with respect to Christmas trees or ornamental by clipping and pruning or a method of site preparation in which all standing material is removed at ground line using a shear blade attached to a large tractor.

Sheet Erosion

It refers to the gradual and uniform removal of the soil surface by water, without forming gullies or rills.

Shelterbelt

A strip of living trees or shrubs maintained mainly to provide shelter from wind, sun and snow or shelterbelt is an extended windbreak of living trees and shrubs that are established and maintained for the protection of farmlands over an area larger than a single farm.

Shelterwood Cutting

Any regeneration cutting in a more or less regular and mature crop, designed to establish a new crop under the protection (overhead or side) of the old, as typically in shelterwood systems, or where the resultant crop will be more or less regular. **Irregular shelterwood system**: Harvest cutting in which opening of canopy is irregular and gradual; generally in groups, with the final cutting often in strips; and the resultant crop considerably uneven-aged and irregular. Much of the harvesting of Canadian old growth conifer stands composed of tolerant species has been a type of irregular shelterwood. **Strip shelterwood**: A shelterwood system in which regeneration cuttings are carried out on fairly wide strips, generally against the prevailing wind, and progress rapidly; regeneration is mainly natural, regeneration interval short, and resultant crop fairly even aged and regular.

Shelterwood Harvest

The removal of trees on the harvest area in a series of two or more cuttings so new seedlings can become established from the seed of older trees. This method produces an even aged forest.

Shifting Cultivation

Found mainly in the tropics, especially in humid and subhumid regions. There are different kinds; for example, where a settlement is permanent, but certain fields are fallowed and cropped alternately ('rotational agriculture'). In others, whole settlements move and clear new land once the old is no longer productive. Also called 'swidden' ('burnt clearing'), used more to designate the social group, or 'slash and burn', so called because of the operations undergone.

Short Shoot

Where there is shoot dimorphism (two kinds of shoots), these are often specialized shoots with greatly compressed internodes, often bearing leaves and usually flowers; not contributing to the overall architecture of the tree. Sometimes called a 'spur'.

Short Day Plant

A plant whose development is affected by photoperiod, in particular where a process (for example, flowering) is promoted if the plant is subjected to day lengths below (night lengths above) a critical length.

Shrinkage

The contraction of wood caused by drying growth stands. Shrinkage values are usually expressed as a percentage of specific dimensions (or the volume) of the wood when green.

Shrub

A woody perennial plant (lives more than I year) that differs from a perennial herb by its woody, persistent stems, and from a tree by its low stature and branches that start from the base.

Shrubland

Shrubland applies to an open or closed **stand** of shrubs up to about 2 m tall.

Sib

A term meaning brother or sister, that is, an individual belonging to the same family. 'Half sibs' have one parent (usually the seed parent) in common. 'Full sibs' have seed and pollen parents in common. Hence, sibs are progeny of the same parents derived from different gametes.

Sidecast

Earth and other material generated by roadbuilding and deposited on the downhill side of the road.

Significance Level

It is the probability that a statistical test will wrongly reject a true hypothesis. Eg. the probability of inferring that there is a difference between treatments when there is, in fact, none. Specifically, a significance level of 5% (or P = 0.05) implies that there is only 1 chance in 20 that the test will wrongly indicate a real difference.

Silage

Silage is the animal feedstuff resulting from preserving succulent green crops under anaerobic conditions in silos by ensilage process.

Silt

A soil component consisting of particles between 0.05 and 0.002 mm in diameter or the soil textural class with particles between 0.05 and 0.002 mm diameter in USDA system or 0.02 to 0.002 mm in the International System.

Silver Thaw

A weather phenomenon in which great quantities of ice collect on trees and other vegetation, often causing much breakage.

Silvics

The study of the life cycle and characteristics of forest trees and stands with particular reference to local factors and environmental requirements.

Silvicultural Decision Model

A computer model or system that permits the simulation and possibly prediction of the interaction of such factors as site class, access, managed stand volume, and logging costs to assist in decision making regarding silvicultural practices in individual stands.

Silvicultural Harvesting Systems

The manner chosen to harvest timber that include considerations which are not limited to species composition, age of stand, site quality, regeneration methods to be used, market conditions, logging costs, terrain, desired residual stand and management objectives. Two basic types exist: Even Age: Systems which are designed to remove most or all of the trees in a single harvest and produce a stand with trees of a single age class. This system usually relies on artificial regeneration to provide future trees. Eg. Clear cutting, thinning, seed tree, shelterwood.

Silvicultural Regime

A series of stand tending (thinning, pruning, etc.) treatments applied after regeneration to achieve a specific stand management objective.

Silvicultural System

A process, following accepted silvicultural principles, whereby forest crops are tended, harvested, and regenerated resulting in the production of crops of distinctive form.

Silviculture

Silviculture is a branch of forestry that deals with the methods of raising and growing trees or silviculture refers to the theory and practice of controlling forest establishment, composition, and growth which requires the hand of people. This tending of the forest is accomplished through the use of some or all of the following silviculture practices like site preparation, reforestation, thinning, fertilization, vegetation management and harvesting. a) Site Preparation: (for artificial regeneration, planting, seeding) may include brush clearing, discing, and burning. b) Reforestation: (to achieve immediate desired species composition and stocking levels). Includes planting, seeding and or providing for natural regeneration through the leaving of seed trees. c) Thinning: (to reduce competition). Includes pre commercial (before trees are large enough to be of commercial value) and commercial (trees that are large enough for commercial utilization either as sawlogs or chips). d) Fertilization: (to enhance site nutrient levels). May be done repeatedly throughout the life of the forest. e) Vegetation Management: (to reduce competition to crop trees). Most economically achieved with selective herbicides but also with manual labor. f) Harvesting: Depending upon which silviculture cutting system is chosen, harvesting may be the end of one cycle (rotation) or it may be just another period in the cycle of an intensely managed forest using a "selection" harvest system.

Silvopastoral System

An agroforestry system that include trees or shrubs and pastures and animals is refered as silvipastoral system.

Silvopastoral

The integration of trees with pasture is called silvipastoral.

Silvopisciculture

It is the practice of growing trees as part of a fish farming enterprise.

Simple Bud

The bud containing either leaf or flower primordia but not both is called as simple bud.

Simple Random Sample

The sample selected in such a way that each and every individual in the population stands an equal chance of being selected is called as simple random sample.

Simulation Model

It is a type of model that is described and investigated (usually with the aid of a computer) so as to imitate the essential features and behaviour of a real system.

Simultaneous Cropping

A form of multiple cropping where two or more crops (or trees) are being grown on the same unit of land at the same time.

Single Seed Descentbreeding Method

A means of advancing populations while preserving genetic variance for later selection. The method uses less labour and space than the backcross and the bulk breeding methods. A single seed is taken from each plant and used to propagate the population. Eventually, pure lines can be generated from existing lines.

Single Tree Selection

An uneven aged reproduction cutting method in which the trees are selected from all dbh classes for harvest or retention based on individual tree merits.

Single Tree Selection Method

A method of regenerating uneven aged stands in which individual trees are removed more or less uniformly throughout the stand.

Singling

The procedure of removing surplus seedlings or young plants from the clusters (clumps) or close planted lines in which they have been sown or transplanted, so as to leave each plant with sufficient space for its future needs is called as singling.

Sink Root

It is a root, other than a tap root, which grows straight downward in the soil.

Sink

Sink refers to a site within a plant to which particular materials are translocated and then deposited or utilized. Sinks obtain the materials they need from sources, which produce or elaborate them. A knowledge of source sink relationships can assist with the understanding of whether the growth and yield of a crop are being driven by demand (for example, millet) or by supply (for example, groundnut).

Site

A tract of land with reasonably uniform soil and climatic factors. An area with the capacity to produce a particular forest or other vegetation because of biological, climatic and soil factors.

Site Capability

The mean annual increment in merchantable volume which can be expected for a forest area, assuming it is fully stocked by one or more species best adapted to the site, at or near rotation age. Expressed in cubic metres per hectare.

Site Class

A measure of the relative productive capacity of a site for the crop or stand under study, based on volume or height (dominant, co dominant, or mean) or the maximum mean annual increment that is attainable at a given age is called as site class or site class refers to any interval into which the site index range is divided for purposes of classification and use.

Site Description

The description of research site with respect to physical and socioeconomic environments and existing cropping systems is called as site description.

Site Index

A measure of forest site quality based on the height (in feet) of the dominant trees at a specified age (usually 50 years. A site index of 95 means that the expected height of the dominant trees at an index age of 50 years would be 95 feet on a particular area of land or an expression of forest site quality based on the height of the dominant trees at a specified age (usually 50 years).

Site Preparation

Disturbance of the forest floor and topsoil to create suitable conditions for natural or artificial regeneration by mechanical, chemical, or by prescribed burning methods or site preparation refers to preparing an area of land for planting, direct seeding or natural reproduction by clearing, chemical vegetation control, burning, disking, bedding, windrowing or raking is called site preparation.

Site Quality

The capacity of an area of land to produce forests or other vegetation. Site quality is influenced by climate, soil type, slope, and biological factors. Site quality is usually expressed by an index based upon the height of a tree if it were allowed to grow free of competition for 50 years (site index of 80 means a tree on that site could grow to 80 feet in 50 years if it has no competition). Site quality is extremely variable within a small area.

Site

Site refers to an area of land, especially with reference to its capacity to produce vegetation as a function of environmental factors (climate, soil, biology, etc.).

Size Classes

Tree sizes representing stages in the development of a tree or stand.

Skid Trail/Road

The pathway for logs to be moved from the stump to the landing. These paths are intended to be temporary and only used during harvesting operations.

Skidder

A machine that transports logs from the stump to the landing.

Skidding

The process of dragging logs from the woods to a landing, usually applied to ground-based operations. A similar term, used especially with cable or aerial logging systems, is "yarding." Or it refers to the removal of logs from a logging operation by dragging them along the ground attached by ropes or chains to a power source, such as horses or mechanical tractors.

Skyline Logging

A type of cable logging in which the mainline is stationary and a carriage moves up and down it, collecting turns of logs.

Slab

The exterior portion of a log removed in sawing timber.

Slash

In forestry, the vegetation (branches and other woody and leafy debris) left on the forest floor after trees have been felled or trimmed is called as slash.

Slash and Burn System

A kind of shifting cultivation practiced in high rainfall areas where the cropping period is followed by a fallow period during which grass, herb, bush or tree growth occurs or it can also defined as a pattern of agriculture in which existing vegetation is cut, stacked and burned to provide space and nutrients for cropping; also called 'swidden' cultivation and shifting cultivation.

Slashing

The practice of cutting back the less tough, competing vegetation, for example, ground cover like bracken. A form of clearing.

Slope

Slope is a measure of steepness, or the degree of deviation of a surface from the horizontal, measured as a numerical ratio, percent or in degrees.

Slurry

Slurry is the semi liquid animal waste (faeces and urine) that accumulates where animals are confined with little or no bedding. It can be diluted and spread on grassland or crop fields, but there is a risk of spreading salmonella infection.

Small Roundwood

Normally refers to logs of 7 cm maximum top diameter, over bark is called small roundwood.

Smallholder

A farmer who is relatively resource poor, who cultivates or keeps animals, or both, on only a small piece of land, sometimes only a small **plot**. These farmers may or may not have access to other common land.

Snag

Snag refers to any dead or dying tree suitable as a nest or roost site for a cavity nesting animal. Most birds prefer snags that are at least 10 inches DBH, and the larger, the better.

Social Forestry

Refers to the planting and tending of trees or shrubs for the well being and betterment of local communities.

Socio Economic Determinants

The factors such as marketing facilities, the land tenure system and credit, which influence the cropping systems of a given area are called as socio economic determinants.

Sod Seeding

It refers to the practice of sowing directly into a **sward** (grass or grass plus weeds) without prior cultivation.

Sod

Sod is a surface layer of matted vegetation which is held together by the roots, rhizomes and stolons of grasses and other herbs.

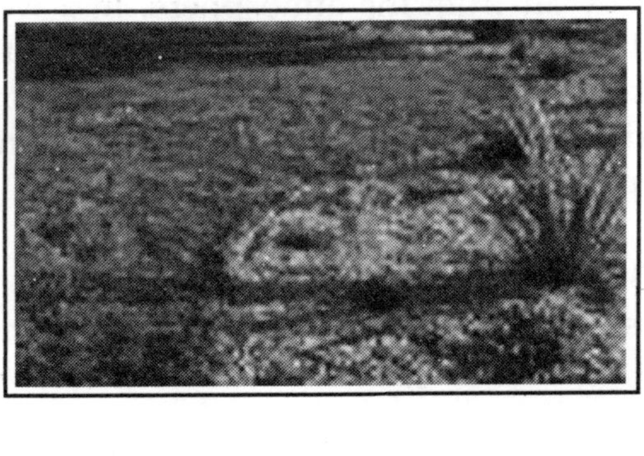

Softwood

A term used in the timber trade to describe the wood of most conifers (gymnosperms), as distinct from the hardwood, broadleaved species (angiosperms).

Softwoods

A tree belonging to the order Coniferales, usually evergreen, cone bearing and with needles or scale like leaves such as pine, spruces, firs, and cedars; often called softwoods. Bald cypress is a deciduous conifer.

Soil Aeration

Soil aeration is the process by which air in the soil is replenished by air from the atmosphere. In a well-aerated soil, the soil air is similar in composition to the atmosphere above the soil. Poorly aerated soils contain more carbon dioxide and less oxygen. The degree of aeration depends on the soil porosity.

Soil Amendment

It refers to an any material incorporated into a soil to improve its fertility, long term chemical or physical status, or stability, such as fertilizers (including slow-release fertilizers, trace elements), organic materials (dung, farmyard manure, compost, straw, bark), sand or peat, lime, gypsum, synthetic resins that improve water-holding capacity, and so on.

Soil Bulk Density

The ratio of dry weight of a soil mass to its volume is called soil bulk density.

Soil Class

A group of soils having a definite range in a particular property such as acidity, degree of slope, texture, structure, land capability, degree of erosion, or drainage.

Soil Classification

The systematic arrangement of soils into groups or categories on the basis of their characteristics is called soil classification.

Soil Conservation

It is a combination of all management and land use methods that safeguard the soil against depletion or deterioration (loss of fertility) caused by nature or humans.

Soil Dry Weight

The weight of solid soil particles when all water has been vapourized after heating the particles to a temperature of 105°C is called soil dry weight.

Soil Fertility

Soil fertility refers to the quality that enables a soil to provide adequate nutrients in a proper balance for specified plant growth when other factors such as light, moisture, temperature and the physical condition of the soil being favourable.

Soil Horizon

A layer of soil with distinct characteristics that separate it from other soil layers. Commonly, a forest soil will have 0, A, B, and C horizons.

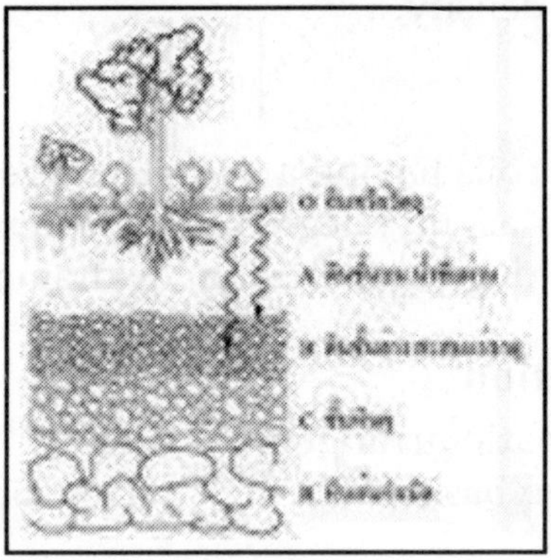

Soil Horizon

It is a layer of soil approximately parallel to the land surface and that differs from adjacent layers in physical, chemical and biological properties or characteristics such as colour, structure, texture, consistency, kinds and numbers of organisms, degree of acidity. Soil may consist of an **'O horizon'** (of **litter**); an **'A horizon'** (darkened by organic matter, but from which clay or sesquioxides have been lost); a **'B horizon' (subsoil,** with natural ground from the A horizon); and a **'C horizon'** (little affected by the biological activities).

Soil Moisture

The relative amount of water in the soil; usually applied to upper levels of soil, occasionally to humus layer.

Soil Moisture Deficit

It is the difference between the existing soil water status and that which would obtain at field capacity.

Soil Organic Matter

Material found in soil derived from living matter is called soil organic matter. 'Fibric' organic matter is the least decomposed and is mainly fibres; it has low bulk density and a fibre content > 60% of the organic volume. 'Hemic' organic matter is intermediately decomposed; 'sapric' is the most decomposed, with the highest bulk density and the least fibre. Other ways of fractionating soil organic matter (such as into 'labile' and 'stable' fractions) are based on treatment with chemical agents, for example, different strength solutions of potassium permanganate.

Soil Permeability

Soil permeability is the quality of a soil enabling it to pass air or water.

Soil Porosity

Soil porosity is the percentage of the soil (or rock) volume that is not occupied by solid particles, including all pore space filled with air and water. The total porosity includes both 'capillary' and 'non capillary' porosity (capillary refers to the very small pores).

Soil Profile

A vertical section of soil showing the nature and thickness of the various horizons, often used in soil classification or soil profile refers to a section of two dimensions extending vertically from the earth's surface so as to expose all the soil horizons and a part of the relatively unaltered underlying material.

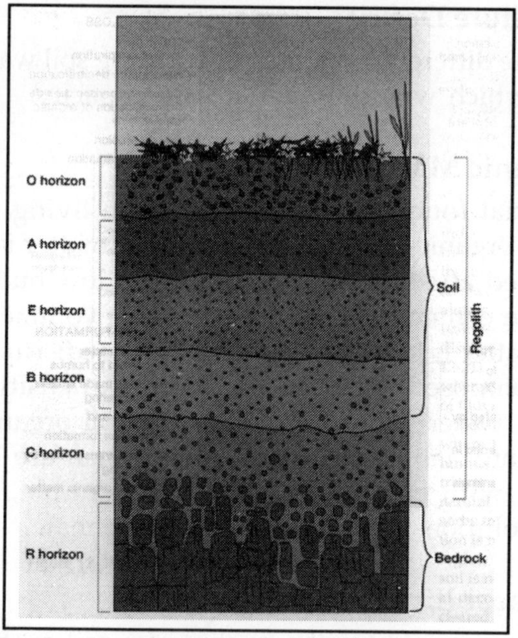

Soil Reaction

The degree of acidity or alkalinity is called as soil reaction.

Soil Series

Groupings of soils with similar profile characteristics.

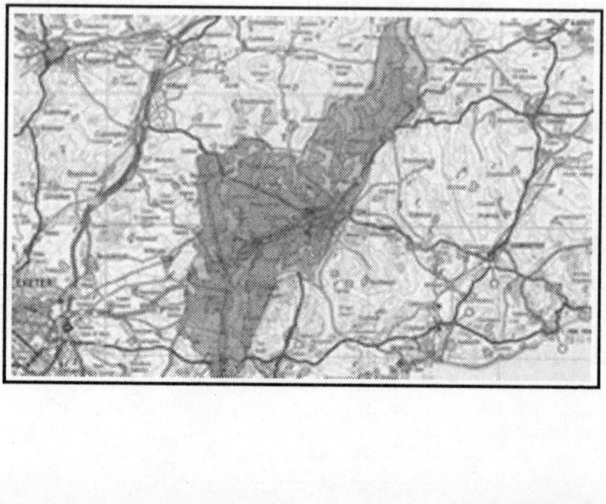

Soil

Soil is a dynamic natural body composed of mineral and organic materials and living forms in which plants grow. The collection of natural bodies occupying parts of the earth's surface that support plants and that have properties caused by the integrated effect of climate and living matter acting upon parent material, as conditioned by relief, over periods of time.

Soil Structure

It is the combination and arrangement of primary soil particles, units or peds. The secondary units are classified on the basis of size, shape and degree of distinctness, such as 'crumbly', 'angular', 'granular', 'platey', 'prismatic', 'columnar', 'blocky'.

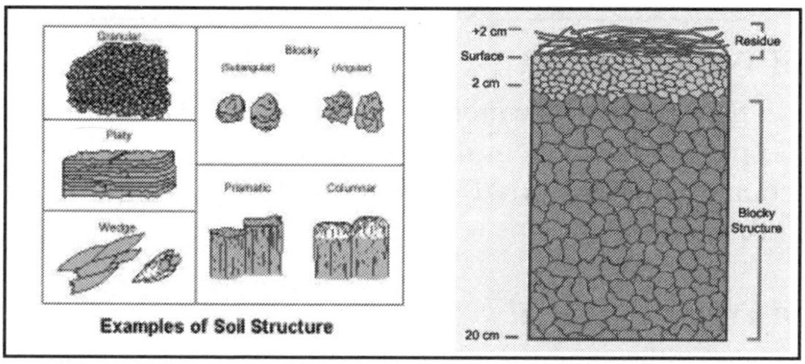

Soil Survey

The systematic examination, description, classification and mapping of soils in an area is called soil survey.

Soil Texture

Soil texture refers to the relative proportion of the various size groups of individual soil particles. 'Fine fraction' is silt and clay-sized particles < 0.05 mm; 'coarse fraction' is the stone, gravel and sand > 0.05 mm.

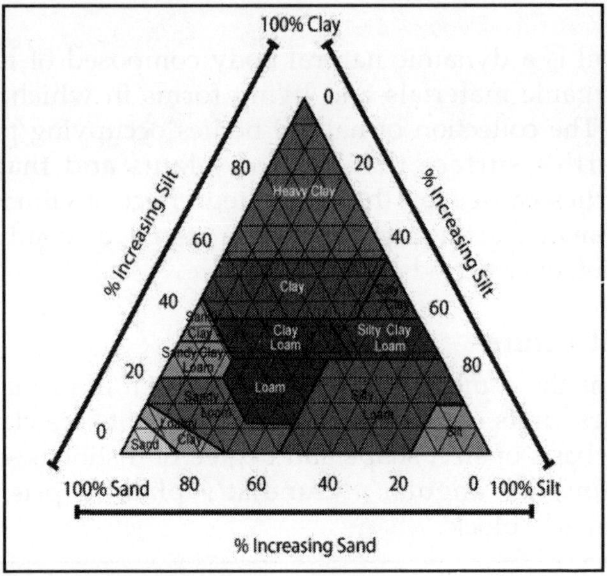

Soil Water Content

The quantity of water in a soil to a stated depth. It can be expressed as a weight fraction (g g–1) or a volume fraction (g ml–1), so that volume fraction = weight fraction x bulk density.

Soil Water Potential

It is defined as the amount of work that must be done per unit quantity of water to transport irreversibly and isothermically an infinitesimal quantity of water, at a specified elevation and at atmospheric pressure to the soil (at the point under consideration).

Soil Water Storage

The water retained in the soil, not draining. Soils with a higher proportion of clays usually retain more water than those with a high content of sand, although it may be less available to plants.

Sole Cropping

Crop cultivar grown in pure stands at normal planting density— monoculture. Synonymous with 'solid planting' and the opposite of intercropping.

Solifluction Soils

Soils found in high altitude areas where alternate thawing and freezing causes the soil surface to heave and move. It is difficult for seedlings to establish on such soils.

Solifluction

Solifluction refers to the slow flow of material on sloping ground, characteristic of, though not confined to regions subjected to alternate periods of freezing and thawing.

Solum

Solum refers to the upper and most weathered part of the soil profile; the A and B horizons. (Plural is 'sola').

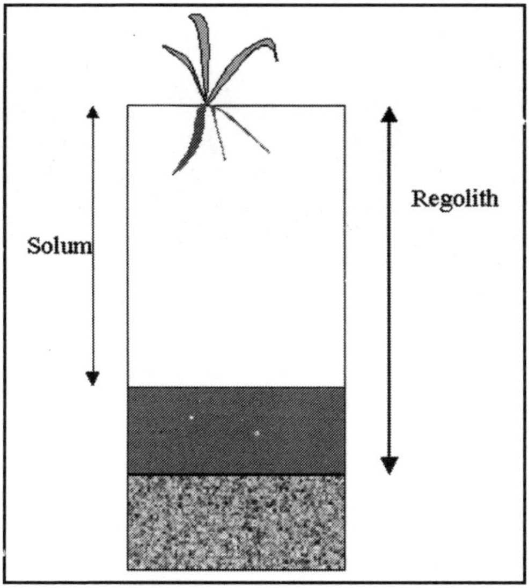

Source

Source is a state variable outside the system's boundary, which is not quantified and from which inputs are derived whereas in crop physiology, a location within a plant that derives and provides particular substances that are used or deposited elsewhere (to sinks). For example, photosynthethizing leaves are a 'source' for carbon assimilates (carbohydrates) required by developing fruits.

Source Identified Seed

The seed attested by a recognized certifying agency as being from the specific seed source (syn: 'standard seed') is called source identified seed.

Spacing

The distance between trees in a plantation, a thinned stand, or a natural stand.

Spar

A pole, tower, or tree used in cable logging to raise the mainline off the ground.

Species (of trees)

The term applies to trees having very similar genetic makeup, so that they freely interbreed and have common characteristics. In common language, a "kind" or "variety." Each species is identified by a scientific name that consists of a genus portion and then a species portion (Tsuga heterophylla, western) or species is defined as a group of related organisms having common characteristics. Ponderosa pine and sugar pine are the common names of two different species.

Specific Combining ability

Specific combining ability (SCA) is the ability of the progeny from a particular full sib cross to perform better than what would be predicted from the general combining ability of either parent.

Specific Leaf Area

The area of a leaf per unit leaf weight which is expressed as square decimetres per gram.

Specific Leaf Volume

The volume of a leaf per unit weight of that leaf.

Spike Top

A tree with a dead top, usually a mark of declining vigor.

Split Plots

A form of layout for experiments in which levels of different factors are applied to different sized plots (for example 'plots' and 'subplots').

Springwood

The less dense, larger celled, first-formed wood of an annual growth ring is called springwood.

Sprout

A shoot from a dormant bud, often at the base of a tree, or from an exposed root or stump (syn: root sucker, 'stump sprout'). A stump with one or more sprouts is called a stool or a young tree developed directly from the base, stump, or root of another tree. Relatively common among hardwoods; with conifers, typical only of redwoods.

Spud

A hand tool which is used for stripping bark from felled trees.

Spur

A short shoot is called as spur.

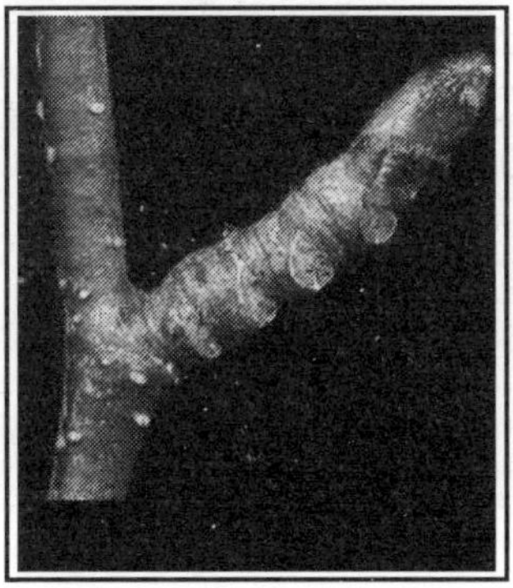

Stag Head

The tree dead at the top as a result of injury, disease, or deficient moisture or nutrients is called as stag head.

Stagnant

Stagnana refers to a stand whose growth and development have all but ceased due to poor site and/or excessive stocking.

Stakeholder

Concerned individuals or groups.

Stand

An aggregation of trees occupying a specific area and uniform enough in composition (species), age, and arrangement to be distinquishable from the forest on adjoining areas.

Stand (Mixed)

A stand in which more than 25% of the trees in the main crown canopy are of a species other than the major species.

Stand (Pure)

A stand in which at least 80-90% of the trees in the dominant and codominant crown classes are of a single species.

Stand

A community of trees possessing sufficient uniformity in composition, age, arrangement, or condition to be distinguishable from the forest or other growth on adjoining areas, thus forming a silvicultural or management entity or a group of forest trees of sufficiently uniform species composition, age, and condition to be considered a homogeneous unit for management purposes.

Stand Density

A quantitative measure of tree cover on an area in terms of biomass, crown closure, number of trees, basal area, volume, or weight. In this context, *tree cover* includes seedlings and saplings; hence, the concept carries no connotation of a particular age. Expressed on a per hectare basis.

Stand Density Index

Any index for evaluating stand density such as those of Curtis, Mulloy, Reinecke.

Stand Density Management Diagram

A two dimensional graph showing the logarithmic relationship between declining mean stand frequency and increasing mean tree size as mean stand diameter and dominant height increase in pure even-aged stands.

Stand Model

A computer model that forecasts the development of a forest stand, usually in terms of mean stand attributes, e.g., mean diameter, height.

Stand Table

A summary table showing the number of trees per unit area by species and diameter classes, for a stand or type. The data may also be presented in the form of a frequency distribution of diameter classes.

Standard

A tree selected to remain standing, after the rest of the stand has been felled over a younger or a new crop, for some special purpose, e.g., shelter, seeding, production of a special quality or size of timber.

Standard Deviation

A commonly used measure of the variability of a set of data; the square root of the variance of a sample.

Standard Error

Standard error is the measure of the error in an estimated quantity (mean, correlation or regression coefficient), defined as its standard deviation.

Standing Ground

A sheltered site with facilities for maintaining young nursery stock (often being grown in containers) prior to the plants being planted out in their final position.

Stands, Fully Stocked

A forest stand with all the growing space effectively occupied but having ample room for the developing crop trees.

Stands, Old Field Stand

A stand on land once used for agricultural crops or for pasture.

Stands, Overstocked

Overcrowding in a stand leading to retarded growth, mortality and increased fire danger.

Stands, Pole Timberstand

A stand where most trees are from 5 to 10 inches in diameter.

Stands, Residual Stand

The stand remaining after cutting.

Stands, Sawtimber Stand

A stand where most trees are large enough in diameter (usually 10 to 12 inches DBH or larger) to be sawed into lumber.

Stands, Start Date of Contract

The date upon which the timber harvesting may begin. It is usually stated in the contract.

Stands, Understocked

A stand with the growing space not effectively occupied by crop trees.

State Variable

In simulation modelling, a variable that characterizes the state of the system and directly determines the processes that bring about changes in the endogenous variables.

Statistic

Any quantity calculated from data is called as statistic.

Stem

An above ground axis (usually) of a plant, which develops from the epicotyl (seedling stem tissue above the cotelydon) of the embryos, or from a bud of an already existing stem or root or it can also defined as the principal axis of a plant, carrying all the accessory parts such as the branches, leaves and flowers.

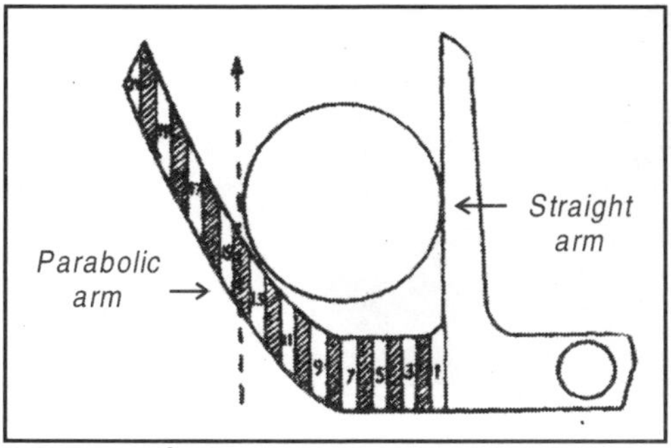

Finnish parabolic calliper

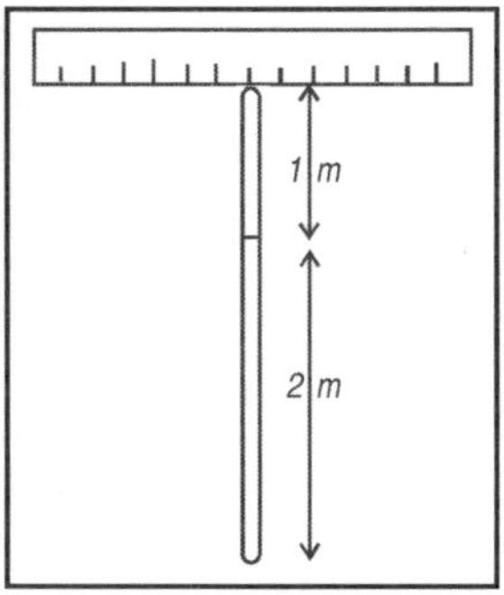

Ruler for measuring stem diameter

Stem Pitting

Stem pitting is a symptom of viral diseases characterized by depressions on the stem of the plant.

Stemflow

Rainfall captured by the above ground parts of a tree that flows down the trunk (stem) to the ground (that is, that which is not retained in the bark, which is the 'bark storage factor') is called stemflow.

Sterility

The inability to produce sound (viable) seeds is called as sterility.

Stewardship

Practices that relate to the care of the land. Usually implies good practices.

Stochastic

A stochastic situation is one in which a given input leads to a number of possible outputs, each with a probability of occurrence.

Stochastic Model

A model in which one or more of the functional relations depend on chance parameters and are hence related to a probability distribution.

Stock Plant

A plant kept to conserve a selected genotype, which can be multiplied by one form of vegetative propagation or another. Also known as 'nursery stock'.

Stock

Stock is a part of a plant, usually consisting of the root system, together with a part of the stem, onto which is grafted a part of another plant.

Stock Table

A summary table showing the volume of trees per unit area by species and diameter classes, for a stand or type.

Stocking

An expression of the adequacy of tree cover on an area in terms of crown cover, per cent stocking, number of trees, basal area or volume in relation to a pre-established norm (optimum for growths). **fully stocked**: Productive forest land stocked with trees of merchantable species. These trees by number and distribution or by average dbh, basal area, or volume are such that at rotation age they will produce a timber stand that occupies the potentially productive ground. They will provide a merchantable timber yield according to the potential of the land. The stocking, number of trees, and distribution required to achieve this will be determined from regional or local yield tables or by some other appropriate method. **nonstocked**: Productive forest land that lacks trees completely or that is so deficient in trees, either young or old, that at the end of one rotation, the residual stand of merchantable tree species, if any, will be insufficient to allow utilization in an economic operation. **normally stocked**: Productive forest land covered with trees of merchantable species of any age. These trees, by number and distribution, or by average dbh, basal area, or volume, are such that at rotation age they will produce a timber stand of the maximum merchantable timber yield. This yield must satisfy the site potential of the land as reported by the best available regional or local yield tables. For stands of less than rotation age, a range of stocking classes both above and below normal may be predicted to approach and produce a normal stocking at rotation age and may, therefore, be included. This is because greater or lesser mortality rates will occur in over- or understocked stands as compared with those in a normal stand. **NSR** (not sufficiently or not satisfactorily restocked or regenerated): Inadequate stocking. Productive forest land that has been denuded and has failed partially or completely to regenerate naturally or to be artificially regenerated. The regeneration must contain a minimum number of well established, healthy trees that are free to grow and sufficient to produce a merchantable timber stand at rotation age. In a number of provinces, forest lands that have been recently harvested and that are not yet classified as satisfactorily stocked are classified as NSR. This classification does not necessarily imply that re-treatment is required to correct a failure to regenerate. **overstocked**: Productive forest land stocked with more trees of merchantable species than normal or full stocking would require. Growth is in some respect retarded and the full number of trees will not reach merchantable size by rotation age according to the regional or local yield or stock tables for the particular site and species. **partially stocked**: Productive forest land stocked with trees of merchantable species insufficient to utilise the complete potential of the land for growth such that they will not occupy the whole site by rotation age without additional stocking. Explicit definition in stems per hectare, crown closure, relative basal area, etc. is locally or regionally defined and is site specific. **satisfactorily stocked**: Productive forest land that has been regenerated naturally or artificially to at least a minimum number of well established, healthy trees of merchantable species that are free to grow and sufficient to produce a merchantable timber stand at rotation age.

Stocking

An indication of growing-space occupancy relative to a preestablished standard. Stands are often classified as understocked, well stocked, or overstocked.

Stocking Rate

It refers to a number of grazing or browsing animals kept on a unit area of land. Usually expressed as livestock units per unit area. It applies to the potential for maintaining animals; the actual number or the livestock units actually being grazed at any time is the 'stocking density'.

Stumpage

Stumpage refers to the value of timber as it stands uncut in the woods and in a general sense, the standing timber itself.

Stock Scion Interaction

It refers to the phenotypic effects of the rootstock on the scion (or occasionally vice versa). Scions perform differently when on one rootstock than if on another rootstock or on their own roots.

Stolon

A horizontal stem at or below ground surface that gives rise to a new plant at its tip. Less commonly, a shoot that bends to the ground and takes root or an above ground prostrate (lying on the surface) stem; may form roots at the nodes that come into contact with the ground.

Stool

In silviculture it refers to a living stump capable of producing sprouts, whereas in propagation it can be defined as a living stump maintained to produce cuttings, layers, etc.

Stopping

Stopping refers to the process of removal or suppression of apical buds to stimulate the production of lateral branches.

Story

A horizontal stratum or layer in a plant community; in forests, appearing as one or more canopies. A forest having more than two storys is called **multistoried.** A forest having one story (the main story) is called single storied. A forest having two storys (the overstory and the understory) is called two storied.

Stover

Stover is the air dried stalks of cereals after removal of the grain.

Strain

Strain is a group of plants that are related by common descent, differing in some aspects from the main body of the species. In forestry the exact content of a strain is less well defined than is that of most other subspecific categories.

Strategy

Strategy refers to the mechanisms by which an organism's genotype is preserved.

Stratification

Stratification refers to a pregermination treatment to break dormancy in seed and to promote rapid, uniform **germination.** The seeds are exposed to moisture at a temperature just above the freezing point (1–5°C) for a specified time. It also refers to the arrangement of plants in layers, for example, a stratified canopy in a forest.

Stratified

The term is applied to the crown cover arranged in layers corresponding to species or age classes or to the size of a plant as a 'tree layer', 'shrub layer', 'herb layer'.

Stream Crossing

Location where logging or forest management activities require crossing a body of water. It may be accomplished by fording, bridging or low water crossings.

Streamside Management Zone (SMZ)

A strip of land of varying size and shape maintained for protecting a sensitive area such as a stream.

Stress

Stress is the effects on plants of a lack or excess of environmental resources (light, water, nutrients), either singly or combined. 'Water stress' includes soil water and the effects of high saturation water vapour pressure deficits. Unsuitable temperatures (high or low) and frost, as well as the effect of wind, either from direct mechanical damage or as the combined effect of exposure, are also factors.

Strip Cropping

The system of growing two or more crops simultaneously in different bands wide enough to permit independent cultivation but narrow enough for the crops to interact agronomically to prevent wind and soil erosion is called strip cropping or growing crops in a systematic arrangement of strips or bands to serve as vegetative barriers to wind and water erosion or the practice of growing crops in narrow bands along the contour in an attempt to reduce runoff, thereby preventing erosion or conserving moisture.

Strip Cutting

Removal of the crop in strips in one or more operations, generally for encouraging regeneration or protecting fragile sites.

Stripling

Tall planting stock from which the lower leaves have been removed (stripped) to reduce transpiration losses is called as stripling.

Structure

The distribution of trees in a stand or group by age, size, or crown classes (e.g., all aged, even-aged, uneven aged, regular, and irregular structures).

Stubble Mulching

The practice of leaving the 'stubble', that is, the remains of the bottom part of the stalks of a cereal crop (usually after mechanical harvesting), to protect the soil is called stubble mulching.

Stubble

Stubble is the basal portion of certain stiff stemmed plants, particularly cereals, after the top portion has been harvested remaining after browsing or grazing is completed.

Stump Age

The term denotes the number of annual growth rings between the bark and the pith counted at stump height.

Stump Height

The distance from the ground to the top of the stump. Good logging practice dictates that stumps be as low as possible (preferably as low as 12 inches) to reduce waste, to minimize visual impact on the logging site, and to promote resprouting of trees.

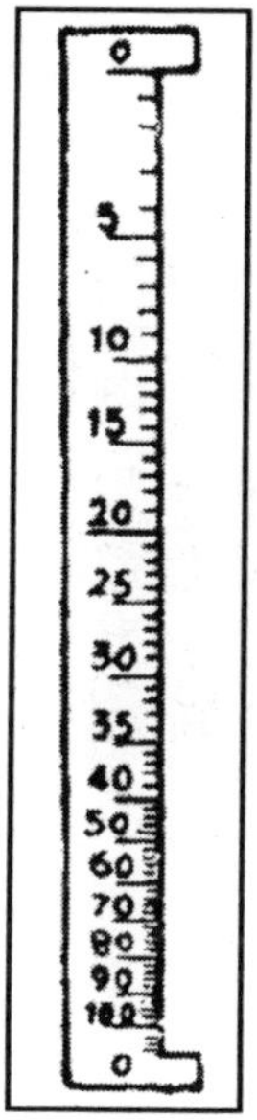

Christens hypsometer

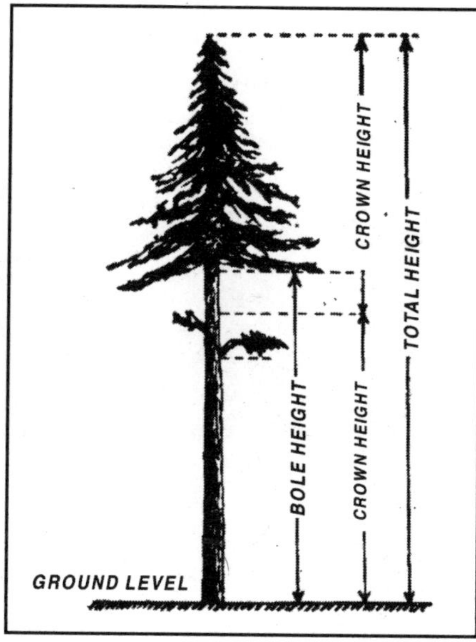

Measurement of height

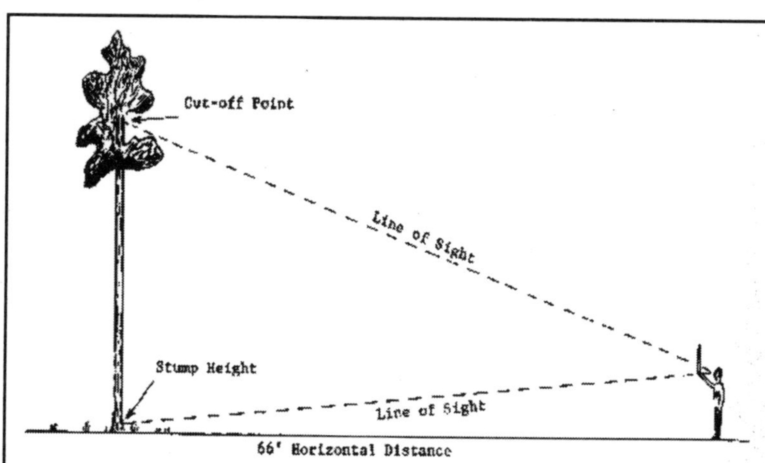

Stump

Stump refers to a planting stock in which the shoot and root have been cut back (usually by some 2–3 cm for the shoot and 10–20 cm for the root) to produce an easily transported propagule.

Stump Treatment

Application of herbicides to or near hardwood stumps to prevent sprouts and suckers.

Stumpage Price

The price paid for standing forest trees.

Stumpage

The monetary value of timber as it stands uncut in the woodlot or the value or volume of a tree or group of trees as they stand in the woods uncut (on the stump).

Sub Compartment

A subdivision of a compartment, in which there is a difference in the tree crop that makes it suitable for different management treatment is called as sub compartment.

Subgenus

Subgenus applies to one of two or more groups of closely related species that together comprise a genus.

Sublining

Dividing a breeding population into several to many smaller populations. All controlled crosses for forward selection are made within a subline, leading to inbreeding within sublines. Production seed orchards are planted with clones or seedlings from several to many sublines to minimize inbreeding in the production orchard.

Subpopulation

A subset of a natural or artificial breeding population. Artificial subsets may be developed for different sites or characters.

Subsistence Farming

The practice of growing crops and, where appropriate, keeping animals so as to provide food (cereals, pulses, vegetable and fruits), shelter materials, and possibly other products (fibres, medicinals) for the use of the family.

Subsoil

The B horizon of soils with weak profile development. Incorrect although common terminology, inherited from the days when soil was seen only as a ploughed layer, defines it as the soil below plough depth in which roots normally grow.

Subspecies

A geographically localized subdivision of a species, genetically and morphologically distinguishable from other subspecies, described according to taxonomic rules, and given a Latin name.

Subsystem

Subsystem refers to any part of a system contributing to the same output as the system itself.

Succession

The natural replacement of one plant (or animal) community by another over time in the absence of disturbance or it refers to the replacement of one plant community by another until ecological stability is achieved. The gradual supplanting of one community of plants by another, the sequence of communities being termed a sere and each stage seral.

Sucker

The term denotes a sprout from the lower portion of a stem, especially from the root.

Sucker knot

A knot associated with a limb growing nearly parallel to the main stem. Sucker knots are not permitted on poles because they funnel water into the pole, promoting decay even if treated. A relative measure of amount of stocking on a forest area, compared with other areas.

Summerwood

The denser, later formed wood of an annual growth ring with smaller cells, with thick cell walls which gives the layer a darker color than that of the springwood. **Sunscald** The death of cambial tissue on one side of a tree, caused by exposure to direct sunlight.

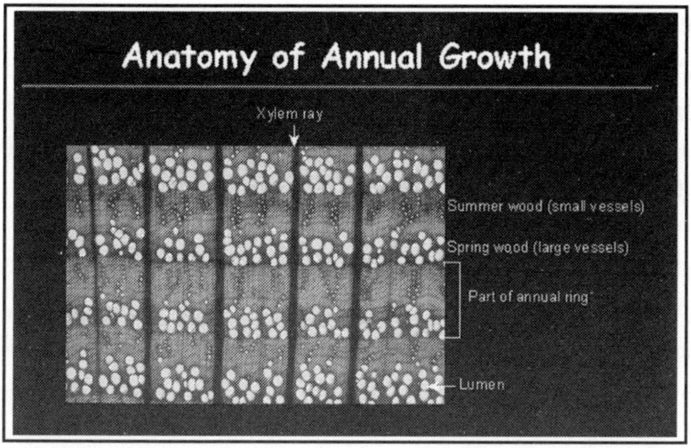

Surface Runoff

Water that moves over the ground surface with the exception of established drainage channels, uncommon on undisturbed forest land.

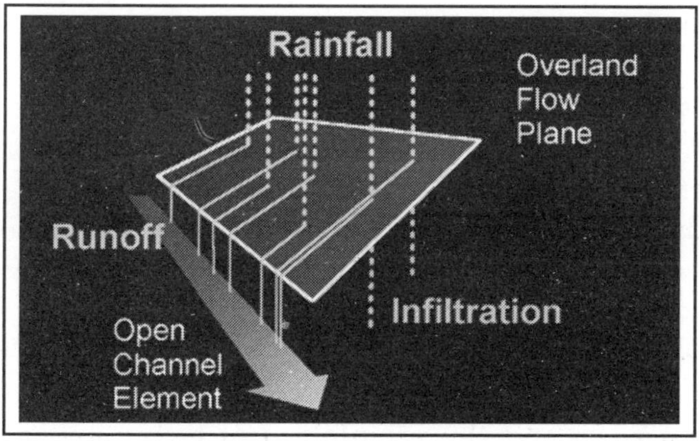

Superimposed Trial

A small set of experimental treatments superimposed on a farmer's production plot, or the fields of a cropping pattern trial at a research site, to evaluate the performance of alternative component technologies usually managed by the researchers.

Superspecies

A taxonomic category comprising two or more species deemed to be sufficiently closely related and of such recent evolutionary divergences to comprise a taxonomic category not far above that of species.

Suppressed

A tree condition characterized by low growth rate and low vigor as a result of competition with overtopping trees. See overtopped.

Surface Active Agent

A substance that, when added to a liquid, affects the physical properties of the liquid surface. Can be used for increasing the wetting properties of sprays and for formulating emulsifiable concentrates and wettable powders.

Survey

A systematic method or an investigation (for example, of farms or individuals) in which the sampling procedure and the measurement process (for example, questionnaire) are fixed in advance and are the same for the whole sample.

Sustainable Development

It refers to the management and conservation of the natural base, and the orientation of technological and institutional change, in such a manner as to ensure the attainment and continued satisfaction of human needs for present and future generations. It conserves land, water, plant and animal genetic resources, is environmentally non degrading, technically appropriate, economically feasible and socially acceptable.

Sustainable Land Use

The term refers to land use that achieves production sufficient to meet the needs of present and future populations while conserving or enhancing the land resources on which that production depends.

Sustained Yield

Management of a forest property for continuous production with the goal of achieving, at the earliest practical time, a balance between growth and harvest or the annual volume of wood products that a forest can produce continuously under a given system is refered as sustainable yield.

Swamp

To clear the ground of underbrush, fallen trees, and other obstructions, to facilitate such later operations as logging or surveying.

Sweep

Sweep is a gradual (but pronounced) bend in a log, pole, or piling which is considered a defect.

Sward

Sward is a grassed area composed of short grasses giving a continuous cover, with no trees or shrubs present.

Swell-butted

Describes a tree greatly enlarged at the base.

Syllepsis

The development of a lateral branch without a period of dormancy, that is, contemporaneous with its parent axis.

Sylvopastoral System

A land use system in agroforestry for the concurrent production of trees and animals that graze or browse or both.

Symbiosis

Symbiosis is a mutually beneficial relationship between two living organisms of different species, living closely together. Eg. Fungus and algae that form a lichen, or nitrogen-fixing bacteria living in roots are examples of symbiosis.

Symbiotic

It denotes the condition of living together. An association of two different organisms that live attached to each other, or one as a tenant of the other, and contribute to each other's support.

Sympatric

Refers to the situation of occurring in the place; inhabiting the same area and having overlapping distributions.

Synecology

The study of groups of organisms that are associated as a unit. The emphasis may be on the fit of one organism into the system, on the relationships between organisms, or on the whole system as a unit.

Synergism

The concurrent parasitism of a host by two pathogens in which the symptoms or other effects produced are of greater magnitude than the sum of the effects of each pathogen acting alone is called as synergism.

Synergistic

The condition of working together or cooperating. In biology, where one organism (or organ) acts with another (or others) to achieve greater results than the sum of their efforts as individuals.

Synthetic Variety

A variety that is produced by crossing among all of a number of genotypes selected for good combining ability in all possible hybrid combinations, with subsequent maintenance of the variety by open pollination is called synthetic variety.

System

System is an arrangement of components (or subsystems) that process inputs into outputs. Each system consists of boundaries, components, interactions between components, inputs and outputs.

Systematic Design

An experimental design laid out with no randomization of treatment plots within a replication. Eg. a fan design or a fertilizer trial in which adjacent plots have gradually increasing spacing or rates of application.

Systematic Sampling

It refers to a sample consisting of units selected in conformity with some regular pattern. Eg. the sample formed from every 10th tree in a row or from the intersections of a regular grid.

Systeme Internationale

An internationally agreed set of units and procedures consisting of 7 base units (metre, second, kilogram, kelvin, ampere, candela, mole), 2 supplementary units (radian, steradian) and derived units. Commonly called SI units.

Systemic

The situation of being found throughout the whole plant. As with soil- or leaf applied herbicides or insecticides that are translocated to all the other parts.

Systems Analysis

The term refers to the orderly and logical organization of data and information into models, followed by the rigorous testing and exploration of the models necessary for their validation and improvement.

Systems Approach

Studying a system as an entity which is made up of all its components and their interrelationships, together with relationships between the system and its environment is called as systems approach.

—xxx—

T

Tail tree

In skyline logging, a tree used to anchor the mainline away from the landing.

Tally

The count of trees, logs, or other products; to count trees, logs, or other products; to record products, distances, etc., as measured.

Tannin

An astringent substance, available naturally in the bark (and sometimes the wood) of different species of trees (for example, acacia and mangrove species); used in converting hide to leather.

Taper

The decrease in thickness, generally in terms of diameter, of a tree stem or log from the base up.

Taproot

The major tree root with the greatest tendency to grow downward. Depending on the species and soil conditions, the taproot may or may not be pronounced or show heavy branching.

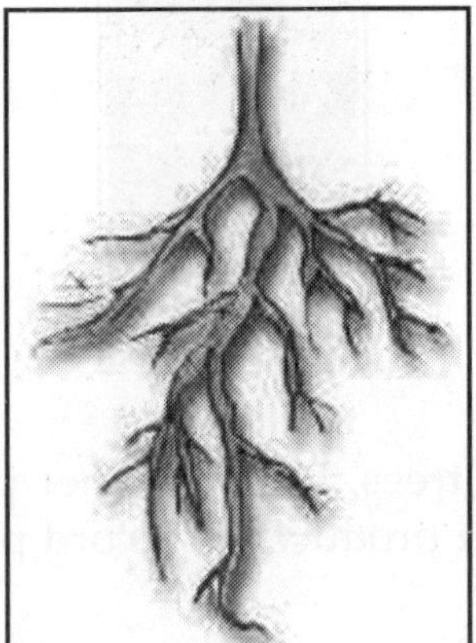

Target Area

Large priority development area in a country for which improved cropping system technologies are developed through on-farm cropping systems research conducted at sites in the area.

Tarif table

A tree-volume table based on d.b.h. and total height.

Taungya System

Method of raising forest trees in combination with (seasonal) agricultural crops. Used in the early stages of establishing a forest plantation. It not only provides some food but can lessen the establishment costs.

Taxon

A taxonomic unit of any size. A category in the taxonomic hierarchy.

Taxonomy

Classification of organisms, including identification and nomenclature, according to a natural (chiefly morphological) system that seeks to depict evolvement. Hence taxon, any designated group within a classification (for example, a cultivar, a species).

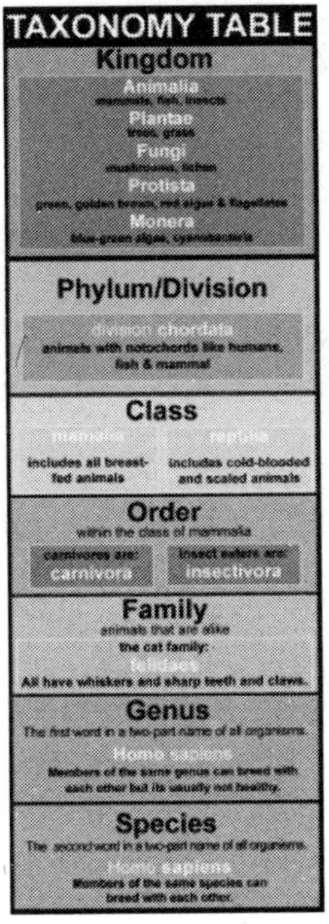

Tending

Generally, any operation carried out for the benefit of a forest crop or an individual thereof, at any stage of its life; covers operations both on the crop itself, e.g., thinnings and improvement cuttings, and on competing vegetation, e.g., weeding, cleaning, and girdling of unwanted growth, but not regeneration cuttings or site preparation. Any operation carried out for the benefit of an established forest crop at any stage of its life (e.g. cleaning, thinning, fertilizing, spraying).

Tendril

A slender, coiling, modified leaf or stem of climbing plants, such as peas, that supports the plant by attaching it to surrounding surfaces.

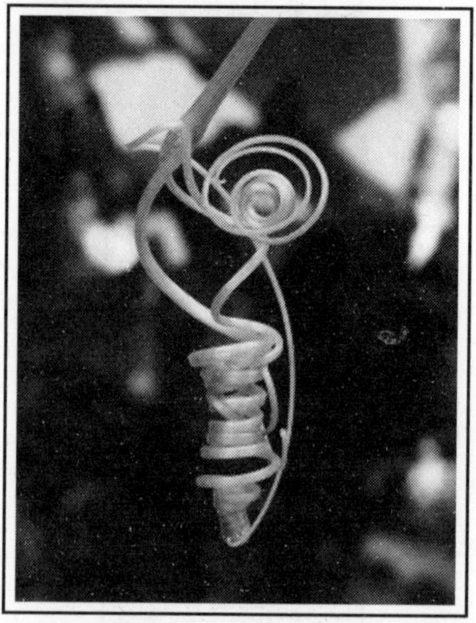

Terminal Bud

A bud that develops at the apex of a stem. Also known as apical bud.

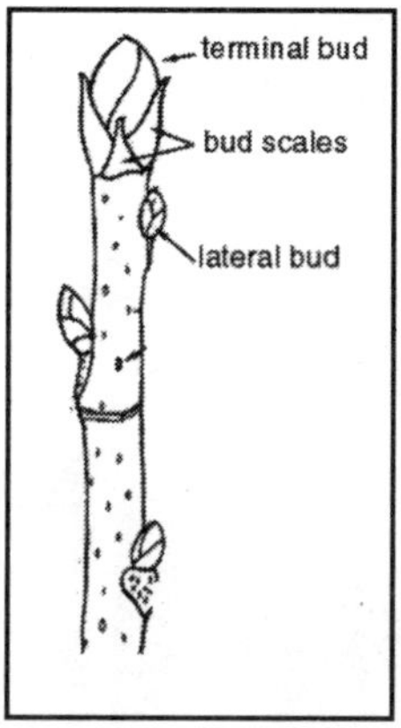

Terminal Bud Scar

A ring of thickened bark on a twig, indicating where the terminal bud grew the previous year.

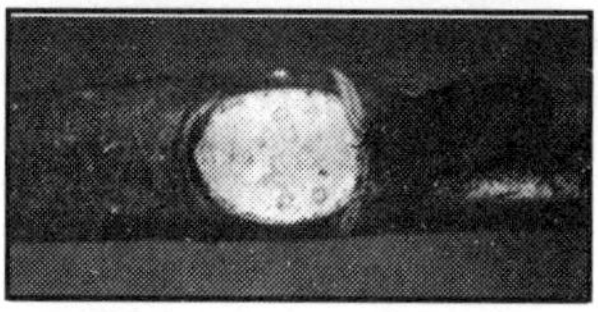

Terrace

A broad surface running along the contour. It can be a natural phenomenon or specially constructed to intercept runoff, thereby preventing erosion or conserving moisture. Terraces are sometimes built to provide adequate rooting depths for plants.

Terrace Cropping

Refers to sowing crops on terraces.

Thermal Time

The time temperature integral above a specified 'base temperature'; usually for a set of processes (for example, growth from sowing to harvest) if individual processes have the same base temperature or it can be defined as the integral of time and temperature in degree days needed for the completion of a particular development process.

Thermograph

Refers to a self recording thermometer.

Therophyte

A self supporting annual (or seasonal) plant whose shoot and root systems die after seed production and which completes its whole life cycle within one year (season).

Thicket

A number of trees or shrubs growing close together or a closed stand of bushes and climbers between 3 and 7 metres tall.

Thinning

A cutting made in an immature crop or stand primarily to accelerate diameter increment but also, by suitable selection, to improve the average form of the trees that remain. (Defined in Nova Scotia as a spacing operation designed primarily to recover potential mortality and to improve growth, quality, and percentage of desirable trees.) **Commercial thinning**: A thinning in which harvested trees are removed from the site and used for commercial purposes. **Crown thinning**: The removal of trees from the dominant and codominant crown classes to favour the best trees of those same crown classes. (1) (Defined in Nova Scotia as the removal of trees from the dominant and codominant crown classes to promote the growth of desirable trees and species.) **Free thinning**: The removal of trees to control stand spacing and favour desired trees using a combination of thinning criteria without regard to crown position. **Low thinning**: The removal of trees from the lower crown classes to favour those in the upper crown classes. **Mechanical thinning:** Thinning involving removal of trees in rows or strips, or by using fixed spacing intervals. **Precommercial thinning (PCT)**: A thinning that does not yield trees of commercial value, usually designed to improve crop spacing. **Row thinning**: A thinning in which the trees are cut out in lines or narrow strips at fixed intervals throughout a stand. **Selection thinning**: The removal of trees in the dominant crown class in order to favour trees in the lower crown classes. **Spacing:** A thinning in which trees at fixed intervals of distance are chosen for retention and all others are cut.

Thinning Cycle

The time interval between thinnings in the same stand.

Thinning Grade

The severity of low thinning based on the crown classes removed, ranging from very light (Grade A) to very heavy (Grade E).

Thinning Intensity

The characteristics of a thinning prescription in terms of the severity and frequency of thinning entries.

Thinning Out

A type of pruning operation in which entire shoots are removed. An extension of, or complement to, heading back.

Thinning Series

Two or more adjacent forest plots that are thinned differently (e.g., to different thinning grades), essentially so as to compare the increment of individual stems.

Thinning Shock

A condition of very slow growth in a thinned stand, usually from a heavy thinning that exposes residual trees to conditions much different from those present before thinning.

Threatened Species

A species or subspecies whose population is so small or is declining so rapidly that it may become endangered in all or a significant portion of its range.

Throughfall

The proportion of the gross precipitation that reaches the ground and is not intercepted by, or retained in, the canopy is called throughfall.

Tie Ridging

In contour furrowing or trenching, a narrow transverse strip of ground left unexcavated (or ridged).

Tillage

Soil-stirring action to loosen soils and so benefit crops by supplying a suitable soil environment for seed germination and root growth and for weed control.

Tiller

A branch that arises at soil level; usually restricted to the grasses or secondary stems that arise from basal axillary or adventitious buds in grasses.

Tilth

The physical state of soil that determines its suitability for plant growth. It takes into account soil structure, soil texture, consistency and pore space. It is a subjective estimation, judged by experience or the physical condition of soil in terms of its ease of tillage, fitness as a seedbed and lack of impedence to seedling emergence and root penetration.

Timber

A term loosely applied to forest stands or their products; often applied to wood in forms suitable for heavy construction (houses, ships, bridges).

Timber Height

The height of a tree to the point at which the stem breaks into branches (the 'crown spring') or to the point at which the diameter of the stem is 7 cm over bark, whichever is reached first. On a felled tree this point is called the 'timber point'.

Timber Stand Improvement (TSI)

A term comprising all intermediate treatments made to improve the composition, structure, condition, and increment of either an even or uneven aged stand.

Tissue Culture

A blanket term for the cultivation of plant tissues in a controlled artificial environment on defined media under aseptic conditions.

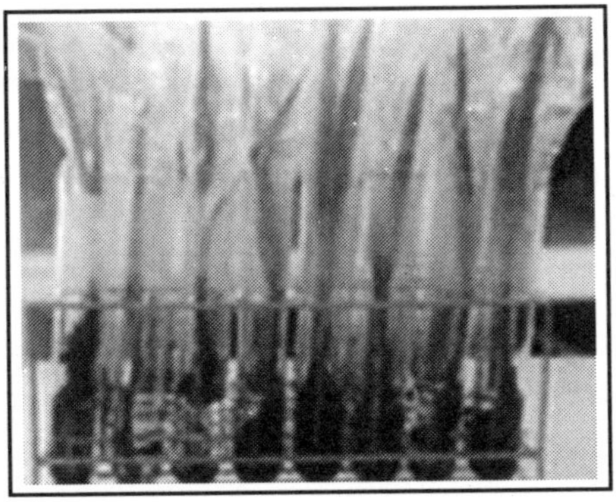

Tolerance

The ability of an organism or biological process to subsist under a given set of environmental conditions. The range of these under which it can subsist, representing its limits of tolerance, is termed its ecological amplitude. For trees, the tolerance of most practical importance is their ability to grow satisfactorily in the shade of and in competition with other trees.

Top Down

Description of research strategies and methodologies in which research planners or officials make programming decisions without the participation or contribution of individuals at other levels of involvement, particularly the target population.

Top Height

In theory, the average height of the hundred trees of largest diameter per hectare. In practice, it is the average height of a sample of the largest diameter trees.

Topoclimate

The climate at 'a place'. Usually considered to encompass larger spatial dimensions than microclimate (for example, the climate around plants), and smaller than macroclimate (for example, of regions), so would cover a 'field' or a 'slope', or similar area.

Topography

A description of the features of a place, or a landscape, especially the rise and fall of the land. On a map it is usually indicated by contour lines.

Topophysis

The long persistence of non-genetic, position or age effects after grafting or rooting. Examples are the horizontal growth of grafted trees produced with **scions** taken from plagiotropic branches and the early flowering of grafted trees produced from flower producing branches.

Toposequence

A sequence of related soils that differ from one another, primarily because of topography as a soil formation factor.

Topsoil

The upper layer of soil rich in organic matter. It forms naturally but is easily lost through disturbance.

Total Age

Total age is the number of years passed since the germination of the seed or the budding of the sprout or root sucker.

Total Area

A spatial unit suggested to embrace agriculture, horticulture, forestry and agroforestry. The complete land area being managed by a single land user (or cooperating group).

Total Forest Age

The average total age of the trees present in a forest.

Toxicity

The degree to which a substance is harmful or poisonous.

Tracheids

An elongate, spindle-shaped xylem cell, lacking protoplasm at maturity.

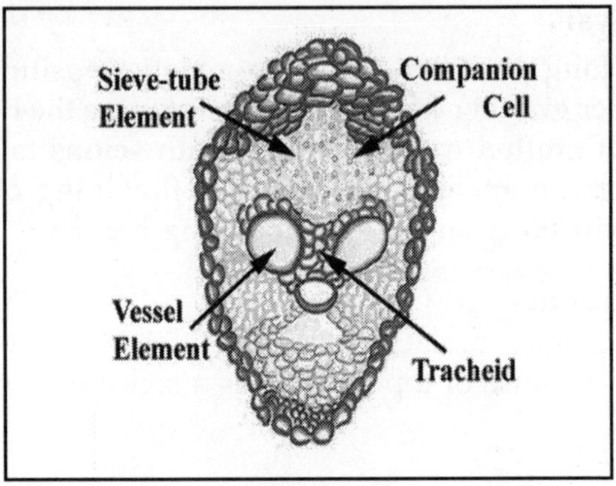

Trainer

A tree beneath the main canopy which by its shading and/or abrasive action hastens the natural pruning or improves the form of some other tree.

Trait

A characteristic of a plant, a group of plants, or part of a plant. Traits may be attributes (quantitative values such as alive/dead or diseased/healthy) or variables (quantitative values such as 'diameter at breast height' or 'basal area').

Transhumance

A situation in which farmers with a permanent place of residence send their herds, tended by herdsmen, for long periods of time to distant grazing areas.

Transition Zone

Of a dicotyledenous stem, the part of the main stem where the anatomy changes from that of a root (vascular strands with internal primary xylem) to that of a stem (with external primary xylem) is called transition zone.

Translocated Herbicide

A herbicide that, after uptake, is moved within the plant and can affect parts of the plant remote from the point of application.

Transpiration

The process by which water vapor leaves a living plant and enters the atmosphere.

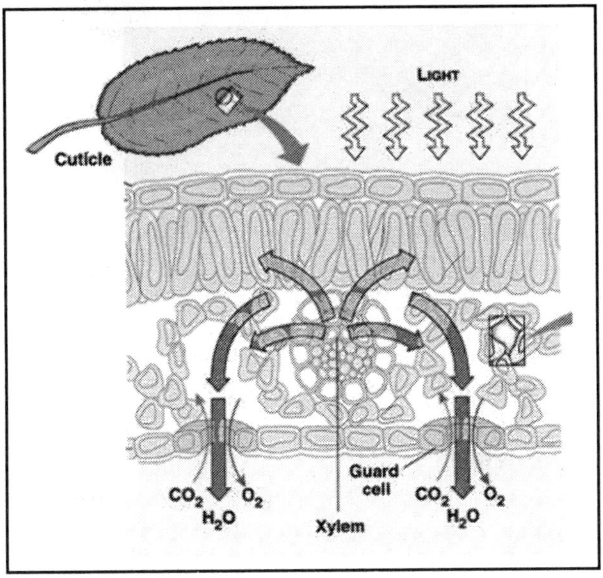

Transplant

A plant that has undergone the process of transplanting, as differentiated from a seedling or cutting, which has not. In a forest nursery, seedlings are normally transplanted after one or two years in the seedbed so as to develop a better root system and a sturdier plant or to take a tree or plant from one location and plant it in another. Transplanting happens in many forms: with seedlings in pots, with open-rooted seedlings, balled-root seedlings and with stumps.

Transplant Lines

Juvenile shrubs and trees planted in nurseries from seedlings or rooted cuttings, to reach acceptable size for planting elsewhere.

Trap Crop

A plant species that provides favourable conditions for certain **pests** and so is grown before or concurrently with but adjacent to the main crop to attract the pests so that they can be suitably eliminated (for example, with insecticides).

Treatment

Any method, procedure, or regime that is applied to some of the experimental units (plots) to compare its effects with those of other treatments. The term includes **control** and control treatment.

Tree

A woody plant having a well-defined stem and more or less definitely formed crown and usually attaining a height of at least 3 m.

Tree Age

The number of years since the germination of the seed, or the budding of the sprout or root sucker.

Tree Farm

An area of privately owned forest land dedicated by its owner to the growing and harvesting of repeated forest crops. The name has been copyrighted by the American Forest Institute for its exclusive use.

Tree Farming

Any agroforestry practice that incorporates trees into farmland.

Tree Form

The degree and mode of taper in a tree or log. Also loosely applied to the general shape of the bole and its desirability for use.

Tree Garden

A multistoreyed agroforestry system in which a mixture of several fruit and other useful trees is cultivated (that is, for a mixture of products), sometimes with the inclusion of annual crops.

Tree Height

The vertical distance between the extreme top of a standing tree (even if this is a lateral shoot) and ground level, using the upper side on slopes.

Tree Length

Entire length of tree, or with the top lopped off at small diameter, as in skidding tree length to a landing for bucking into logs.

Tree Marking

Selection and indication, usually by marking with paint on the stem, of trees to be felled or retained.

Tree Temperament

The set of growth and development reactions shown towards its environment by a tree species during its whole life cycle.

Tree Crop Interface

The spatial extent, above and below ground, over which some form of competition and or facilitation occurs between the tree and the crop components in an agroforestry system.

Trial

An action, method or treatment adopted to ascertain a result. It may be part of a larger 'investigation', and is usually of a shorter or less significant involvement than an 'experiment'. Another word for 'experiment and is often used in applied adaptive field research.

Trickle Irrigation

The process of supplying water to plants using equipment that drips small quantities at a slow rate. Water is usually applied at positions determined by the location of the nozzles although, in some systems (for example, 'seep hose'), it comes from all along the feeder line. Can be above ground or subsurface.

Trim Allowance

The extra 2 or 3 inched left on a bucked log to allow logs with end checks, pulls, or slanting buck cuts to be trimmed to standard lumber lengths.

Truncate

Refers to cut off at the end.

Trunk

The main stem of a tree.

Tuber

A greatly swollen underground shoot or stem, generally irregular in shape, which acts as a store of nutrients or a specialized stem; the enlarged fleshy tip of an underground stem.

Turf

The leafy shoot sytem together with the upper part of the roots cut and lifted (for relaying) from a grassy sward.

Turn

The logs brought to the landing during a single yarding or skidding cycle.

Turnup

In Christmas tree culture, the practice of leaving a green branch when harvesting a Christmas tree; this branch, turned upwards, becomes the next tree.

Twig

Stem one year old or less. Usually applied to a stem not including the leaves

Twiner

Plants that climb by having stem apices that nutate (oscillate and wind) around slender supports.

Two Aged Stand

A stand containing two distinct age classes differing by more than 20% of the rotation age.

—xxx—

U

Under Bark

Measurement of the diameter or circumference of a trunk or branch made with the bark removed.

Underbrush

The brush growing in a forest.

Undercut Seedling

A nursery grown seedling that has had its tap root cut by passing a blade or wire through the seedbed a few centimetres below ground level. This process encourages the growth of lateral roots and is an alternative to transplanting as a means of obtaining sturdy planting stock with a well branched root system. Often moved as balled roots.

Undergrowth

Small trees and shrubs and other plants growing under a forest canopy.

Understory

That portion of the trees or other vegetation in a forest stand below the canopy.

Underplanting

Planting young trees under the canopy of an existing stand.

Understocked

A stand of trees so widely spaced, that even with full growth potential realized, crown closure will not occur.

Understorey

The portion of a tree in a forest stand below the overstorey.

Understory

The level of forest vegetation beneath the canopy.

Underwood

A tree crop growing beneath the main canopy and forming a distinct lower storey. It is usually of coppice but can be of planted or naturally seeded trees.

Uneven-Age

Systems that are designed to maintain a stand with at least three age classes. This system relies solely on natural regeneration to provide future trees. Eg. Selection,thinning, sanitation, salvage.

Uneven Aged

Applied to a stand in which there are considerable differences in the age of the trees and in which three or more age classes are represented.

Uneven-Aged Forest

A forest in which there are considerable differences in age of trees and in which three or more age classes are present.

Uniformity Trial

Planting a test crop at a carefully controlled planting density and under a consistent level of good management over an area on which a field experiment is to be carried out, so as to use the data when it is harvested from small, equal sized plots over the whole area, to assess how 'uniform' this area is. It can be used to define blocks where non uniformity occurs and locational control is part of the experimental design (for example, in a 'randomized block layout', or for covariance analysis).

Unit Sale

A timber sale in which the buyer makes regular (weekly, monthly) payments based on mill receipts. Unit sales are useful when the amount of timber sold is so large that a preharvest, lump sum payment would be prohibitive.

Unmerchantable

Refers to a tree or stand that has not attained sufficient size, quality, and/or volume to make it suitable for harvesting.

—xxx—

V

Validation

The process of assessing the accuracy for a given purpose of a simulation model by comparing the model's predictions with independent results.

Variability

Absence of uniformity. Quality of differing from the average value. This term is usually used in a general sense and qualified by such words as 'low', 'moderate', and 'high'. For quantitative comparisons, the concept of a **variance** is used.

Variable

A quantity able to assume different numerical values. When considering two variables at the same time, one can sometimes be considered the dependent variable and the other the independent variable, for example crop growth (dry matter) and intercepted solar radiation (MJ m–2) respectively or any quantity or quality likely to show variation from one individual to the next in the same population.

Variance

In statistics, of a set of observations, the mean of the squared deviations from their mean value or a statistical measure of variability defined in an exact manner. Many types of variances can be calculated, for example within-block variance and within-plot variance among individual plants. Examining the ratios between two calculated variances provides a way of determining if observed differences, or relationships, are large enough to be caused by factors other than chance.

Variate

An individual observation or value of any variable.

Variety

A taxonomic subdivision of a species based on minor characteristics and often an exclusive geographic range, or an assemblage of cultivated individuals of useful and reproduceable character or in taxonomy, variants of a typical species that are not known to occur as distinct populations but appear sporadically in wild stands as single or isolated individuals.

Vegetative Flowering

The stage of development when a plant is producing only roots, stems and leaves, that is, the vegetative organs.

Vegetative Reproduction

Reproduction by other than sexually produced seed. Includes grafting, reproduction by cuttings, and some types of reproduction without fertilization of seed or seed like organs.

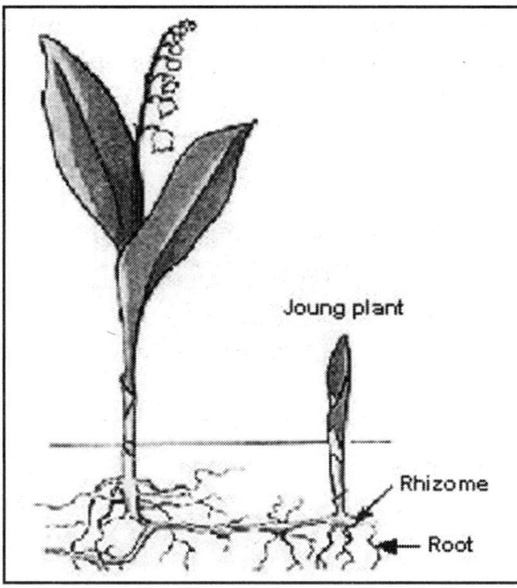

Veld

The name given to grasslands in southern Africa. Can be 'winter veld' or 'summer veld' depending on the climatic pattern, the soil and what grass species occur, hence when it can be grazed. Also spelled 'veldt'.

Veneer

A product made from round logs of certain trees by turning a log and slicing a continuous sheet from it. This sheet is then cut into lengths and glued to lower quality wood for making panelling, table tops, doors and other products with improved appearance.

Veneer Log

A log of suitable size, species, and quality for peeling on a lathe into thin sheets of wood for the manufacture of products such as plywood.

Vernalization

Exposure of a plant or plant organ to chilling, which subsequently brings about a development phase change (for example, flowering).

Vertical Resistance

Resistance to plant pathogens conferred by specific, identified genes.

Vessel

A xylem tube formed from several modified tracheids oriented end to end.

Vexar Tube

A rigid, plastic net tube made from Vexar (a trademark of the DuPont Corporation). Used to protect tree seedlings from animal damage.

Viability

A measure of the potential for seeds to germinate and grow. A seed of short viability will germinate well only within a short time after harvest. Some tree seeds have long viability and will germinate readily, possibly with pretreatment, long after collection or it refers to the possession of life.

Viable Seed

Seeds that will germinate and develop into a plant, given favourable conditions, provided any dormancy that may be present is removed.

Village Forest Garden

Larger than the homegarden, less densely planted and not so well tended. Planted with higher trees and often containing spontaneous species of herbs or lianas.

Virgin Forest

An area of old growth trees that never has been harvested by humans.

Volume Rate

Amount of a liquid applied per unit area. For example, for herbicides: 'high' = > 675 litres ha–1, 'medium' = 226–675 litres ha–1, 'low' = 57–225 litres ha–1, 'very low' = 12–56 litres ha–1, 'ultra-low' = 1–11 litres ha–1.

Volume Table

A table which estimates the volume of wood contained in a standing tree based on measurements of the tree, most commonly dbh and merchantable height.

—xxx—

W

Water Balance

The distribution of water in space, and over time, within a system. Where plants are growing on land this means considering rainfall (and also interception loss, stemflow and throughfall), drainage, evapotranspiration, runoff and soil water storage or it refers to balance of input and output of water within a given defined hydrological area such as a basin or a lake, taking into account net changes of storage. Also used for smaller areas, for example, a field or a plot.

Water Bar

V-shaped trench cut into the surface of a road, which collects water and channels it off the road surface, to avoid erosion.

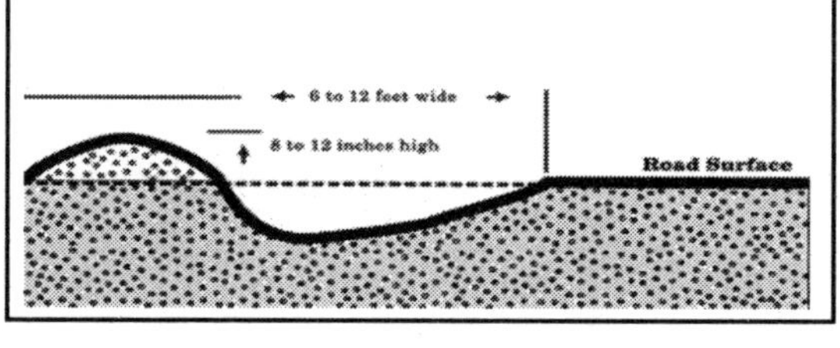

Water Harvesting

A general term for any means of collecting and re-distributing runoff water.

Water Potential

The free energy status of water. A concept of general applicability to the measurement of water in soils, plants and the atmosphere.

Water Spreading

The application of water to lands to increase the growth of natural vegetation or to store it in the ground for subsequent withdrawal by pumps for irrigation or methods of controlling the flow of water locally so as to distribute it better over a farmer's fields. A way of avoiding damage ('gullying') where there is a likelihood of large inflows from any land above a farm or plot.

Water Holding Capacity

Refers to the amount of water that it is capable of holding, between field capacity and permanent wilting point.

Waterlogged

A condition of land where the groundwater stands at a level that is detrimental to plants. It may result from over irrigation, seepage or inadequate drainage.

Watershed

A physiographic unit in the landscape defined by the drainage dividers around the area drained by a particular body of water. If a lake, there is often one watershed with subunits for contributing streams. If a river, it may be defined for any point or all or the whole surface drainage area that contributes water to a lake. The total area above a given point on a stream that contributes water to the flow at that point (syn: 'drainage basin', 'catchment basin', 'river basin') or the total area, regardless of size, above a given point on a waterway that contributes runoff water to the flow at that point. A major drain-area subdivision of a drainage basin on the basis of this concept.

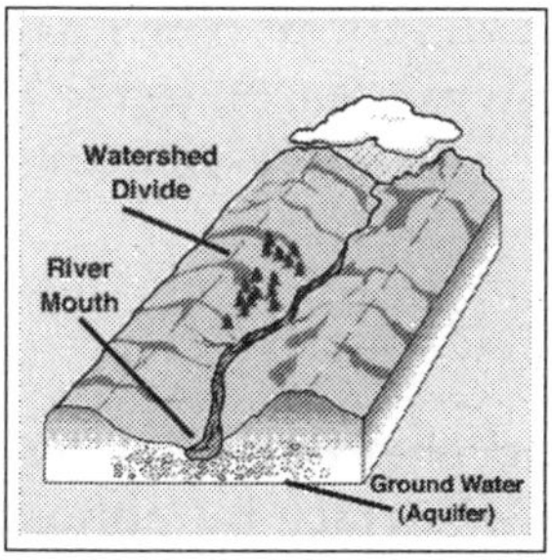

Water Stable Aggregate

A soil aggregate stable to the action of water such as falling drops or agitation as in wet-sieving analysis.

Water Table

The upper edge of free water in the soil. When a hole is dug, water will fill the hole to the level of the water table the level below ground at which free water persists.

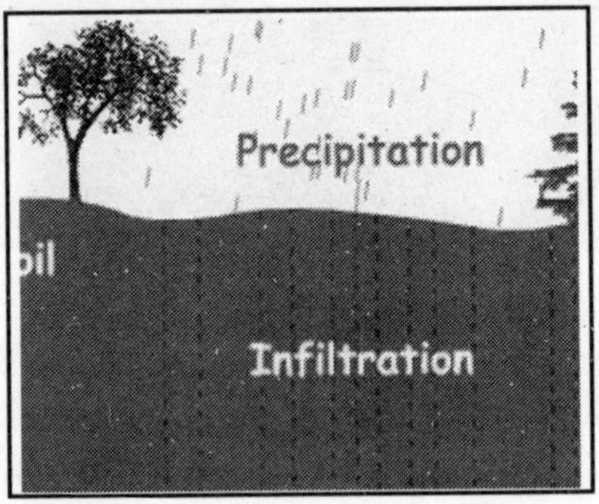

Water Use Efficiency

The above ground dry matter or harvested portion of a crop produced per unit of rainfall received on the same area of ground i.e. including water lost from the soil surface or by runoff and drainage as well as that used in transpiration.

Water Use Ratio

The ratio of the weight of water actually transpired by a crop during its growing season to the weight of dry matter produced (usually exclusive of roots). Also called the 'transpiration ratio'.

Weathering

All physical and chemical changes produced in rocks at or near the earth's surface by atmospheric agents.

Wedge

In logging, to drive a solid V-shaped object (the wedge) into the saw cut to prevent the saw from binding and to direct the fall of the tree or a slang for a prism used in cruising timber.

Wedge System

A modification of the strip shelterwood system in which cuttings begin as narrow, interior, wedge shaped strips with the apex into the prevailing wind, and are then successively enlarged and advanced; regeneration mainly natural; regeneration interval short and young crop fairly even-aged.

Weed

A plant growing where it is not wanted.

Weed Tree

A tree of a species with relatively little or no value. whorl. A group of branches originating at or near the base of the terminal bud. Generally, one whorl is produced each growing season.

Weeding

The elimination of competing vegetation around young trees, for example, by cultivation, by the use of herbicides, or by cutting or slashing. When the trees are big enough to overcome weed competition, a tree plantation is said to be 'established'.

Well Stocked

The situation in which a forest stand contains trees spaced widely enough to prevent competition yet closely enough to utilize the entire site.

Wettable Powder

A type of formulation for pesticides or herbicides for spray application in which the active materials are mixed with an inert carrier, finely ground, and with a surface-acting agent added, so that it will form a suspension when agitated with water.

Wetting Agent

A surface active agent that, when added to a liquid, increases its wetting properties.

Whorl

A ring or circle of plant organs arising close together, for example, leaves from a node, or a spiral of conifer branches from a main stem.

Whorl

A group of branches originating at or near the base of the terminal bud. Generally, one whorl is produced each growing season.

Widowmaker

Any limb, top, leaning tree, or other material in the forest that is in danger of falling to the ground without warning, creating a safety hazard. Often applied to limbs that get lodged in the crowns of other trees during a logging operation.

Wilderness Area

An area of public land (usually over 5,000 acres) designated by Congress as wilderness according to the definition in the Wilderness Act. Ecosystems present are preserved in their natural states, with minimum human influence (no timber harvesting, roads, or developed areas are allowed).

Wildfire

A fire burning out of control, regardless of how or why it started.

Wildling

A seedling naturally reproduced outside of a nursery, used in forest planting.

Windbreak

A wind barrier of living trees and shrubs maintained to protect the farm home, other buildings, garden, orchard, or feedlots.

Wilding

A young seedling that develops in the wild without the help of humans. This type of seedling is the best source of stock for some species.

Wildlife Habitat

The native environment of an animal. Habitats ideally provide all the elements needed for life and growth: food, water, cover and space.

Wildlife Indicator Species

Six wildlife species (moose, woodland caribou, beaver, snowshoe hare, pileated woodpecker, ovenbird) selected to represent the mixture of Habitat types which support almost 250 kinds of forest birds and mammals. Established under the Saskatchewan Forest Habitat Project, each species has a habitat requirement that represents the general needs of a larger group of animals. These measures of biological diversity are important landmarks in the maintenance of healthy and life sustaining forest ecosystems.

Wilting Point

The point in a soil at which the tension of the water it holds causes a plant to start to wilt; when the amount of water held in the fine pores of the soil is insufficient to maintain the turgidity of the plant.

Windbreak

A group of trees or shrubs in any arrangement that will afford protection from high winds to animals or crops or both. When the arrangement is in a long line the group is called a shelterbelt. If an associated reason is also to harvest timber at some future date it is sometimes called a 'timberbelt'.

Windfall

A tree uprooted or broken off by wind; an area on which the trees have been thrown by wind.

Windfirm

Describes trees capable of withstanding heavy wind.

Windrow

Slash, brushwood, etc. concentrated along a line so as to clear the intervening ground between two of them.

Windstrip

A narrow plot of low vegetation (shrubs, bushes, herbs and grasses) that is left when natural vegetation is cleared so as to provide shelter to adjacent crops and to prevent wind erosion on sandy soils.

Windthrow

A tree felled by wind. Windthrows, also known as blowdowns, are common among shallow-rooted species and in areas where cutting has reduced stand density.

Windward Side

The upwind side of, for example, a shelterbelt. Sometimes called the 'weather side'.

Wolf Tree (or Bull Pine)

Usually a tree which is older, larger, or more branchy than other trees in the stand. A tree with a short trunk and a spreading crown which interferes with young growth below.

Wolf tree

A vigorous tree that has merchantable value but occupies more space than its value warrants. Usually very limby.

Wood

Lignified secondarily thickened plant tissue. The structural parts of woody perennials or a small grove of trees of mixed species complete with undergrowth. The size is variable but it does not cover a large area.

Wood Pulp

Mechanically ground or chemically digested wood (composed primarily of wood fibers) which is used in the manufacture of paper, fiberboard, etc.

Wood technology

The study of wood and all its aspects, including anatomy, chemistry, properties, and treatment.

Wooded Grassland

Land covered with grasses and other herbaceous species, and with woody species that cover between approximately 10 to 40% of the ground.

Woodland

An open stand of trees > 8 metres high and with a canopy cover of 40% or more, usually among grasses. Woodland is often described by its dominant species, for example, 'Acacia woodland','Acacia Themeda woodland', 'Combretum woodland'.

Woodlot

A small plot of trees grown for fuelwood, which can also provide small timber and poles.

Wood Processing Industry

The segment of the forest industry that manufactures lumber, paper, plywood, and other primary forest products.

Woody Clump

The aggregate of stems arising from woody organs at or below ground level; initiated from a single individual but sometimes separating to form a group of individual plants of the same genotype.

Woody Perennial

A plant with lignified stems that continues to grow from year to year.

Working Group

An aggregate of forest stands, or forest stand and forest sites, which are grouped for the purposes of applying a common set of silvicultural treatments (also called operational group).

—xxx—

X

Xerophilous

Refers to the ability to withstand drought.

Xerophyte

A plant capable of surviving prolonged periods of moisture deficiency or a plant that normally requires relatively small amounts of water for growth or a plant that grows in extremely dry areas.

Xylem

The principal water conducting tissue or compound vascular tissue and the chief supporting tissue of higher plants; composed of tracheids, vessel members, fibers, and' parenchyma cells'.

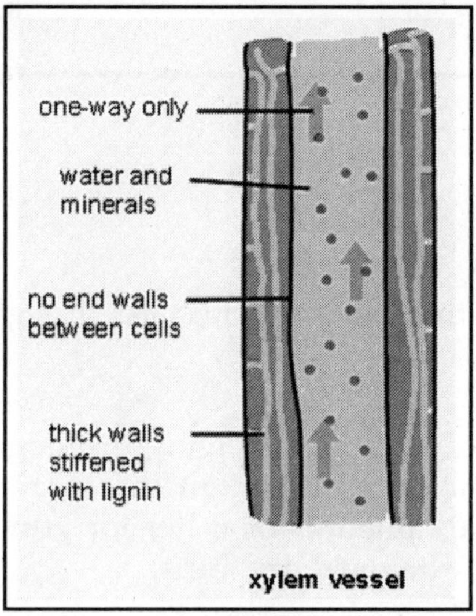

—xxx—

Y

Yield

Plant part harvested and the quantity of it. In agriculture, necessarily related to a specified crop or crops or to a group of animals or to an area, and to a period of time. For multipurpose trees, the amount of each specified part harvested.

Yield Class

The potential or actual yield of a forest, calculated as the total volume of wood produced divided by the age when maximum mean annual increment is achieved. Usually expressed as cubic metres per hectare per year or it can be defined as a way of expressing the productivity of a particular tree crop on a particular site. It is signified by a number that represents the annual production of timber averaged over its economic lifetime or, more correctly, to the age at which the 'mean annual increment' culminates. Thus a crop of 'yield class 10' produces 10 cubic metres of timber per year on average.

—xxx—

Z

Zero-Grazing

A method of keeping animals that involves bringing fodder to them rather than letting the animals graze freely. It is commonly done where land is in short supply. Napier grass *(Pennisetum purpureum)* is a common element in zero grazing systems in East Africa, for example. Trees that can be lopped, like *Leuceana leucocephala,* provide an important addition to these systems. So too do trees like *Acacia tortilis,* which produces edible pods.

Zero Tillage

Growing crops without any significant cultivation of the soil, and often by leaving the previous crop residues on the soil surface as a protective **mulch**.

Zonal Agroforestry

A spatial planting arrangement in which the different species each system remain contiguous to some extent, that is, as strips, plots or even alternate rows; alley cropping is an example. Zonal planting can reduce the tree crop interface compared with mixed planting.

Zonal Soil

A soil type that is defined largely on the basis of climatic belts, where the influence of such factors as parent material is less evident than the influence of climatic development processes.

Zymogenous

Subsisting on easily decomposable organics; responding quickly to freshly added organic materials.

—xxx—